EUROPEAN CONFERENCE OF MINISTERS OF TRANSPORT

CO_2 EMISSIONS FROM TRANSPORT

EUROPEAN CONFERENCE OF MINISTERS OF TRANSPORT (ECMT)

The European Conference of Ministers of Transport (ECMT) is an inter-governmental organisation established by a Protocol signed in Brussels on 17 October 1953. It is a forum in which Ministers responsible for transport, and more specifically the inland transport sector, can co-operate on policy. Within this forum, Ministers can openly discuss current problems and agree upon joint approaches aimed at improving the utilisation and at ensuring the rational development of European transport systems of international importance.

At present, the ECMT's role primarily consists of:
– helping to create an integrated transport system throughout the enlarged Europe that is economically and technically efficient, meets the highest possible safety and environmental standards and takes full account of the social dimension;
– helping also to build a bridge between the European Union and the rest of the continent at a political level.

The Council of the Conference comprises the Ministers of Transport of 36 full Member countries: Austria, Belarus, Belgium, Bosnia-Herzegovina, Bulgaria, Croatia, the Czech Republic, Denmark, Estonia, Finland, France, the Former Yugoslav Republic of Macedonia (F.Y.R.O.M.), Georgia, Germany, Greece, Hungary, Ireland, Italy, Latvia, Lithuania, Luxembourg, Moldova, Netherlands, Norway, Poland, Portugal, Romania, the Russian Federation, the Slovak Republic, Slovenia, Spain, Sweden, Switzerland, Turkey, Ukraine and the United Kingdom. There are five Associate member countries (Australia, Canada, Japan, New Zealand and the United States) and four Observer countries (Albania, Armenia, Azerbaijan and Morocco).

A Committee of Deputies, composed of senior civil servants representing Ministers, prepares proposals for consideration by the Council of Ministers. The Committee is assisted by working groups, each of which has a specific mandate.

The issues currently being studied – on which policy decisions by Ministers will be required – include the development and implementation of a pan-European transport policy; the integration of Central and Eastern European Countries into the European transport market; specific issues relating to transport by rail, road and waterway; combined transport; transport and the environment; the social costs of transport; trends in international transport and infrastructure needs; transport people with mobility handicaps; road safety; traffic management, road traffic information and new communications technologies.

Statistical analyses of trends in traffic and investment are published yearly by the ECMT and provide a clear indication of the situation in the transport sector in different European countries.

As part of its research activities, the ECMT holds regular Symposia, Seminars and Round Tables on transport economics issues. Their conclusions are considered by the competent organs of the Conference under the authority of the Committee of Deputies and serve as a basis for formulating proposals for policy decisions to be submitted to Ministers.

The ECMT's Documentation Service is one of the world's leading centres for transport sector data collection. It maintains the TRANSDOC database, which is available on CD-ROM and accessible via the telecommunications network.

For administrative purposes the ECMT's Secretariat is attached to the Organisation for Economic Co-operation and Development (OECD).

Publié en français sous le titre :
Émissions de CO$_2$ et Transports

Further information about the ECMT is available on Internet at the following address:
http://www.oecd.org/cem/

© ECMT 1997
ECMT Publications are distributed by:
OECD Publications Service,
2, rue André Pascal, F-75775 PARIS CEDEX 16, France.

ACKNOWLEDGEMENTS

The ECMT would like to sincerely thank its Member and Associate Member countries for their assistance in preparing the reports included in this publication. In particular, the national information required for the report "Monitoring of National Policies for the Reduction of CO_2 Emissions" was substantial and required considerable time and effort on the part of Ministry staff. We greatly appreciate the strong support we received. The technical co-operation with experts from OECD, the International Energy Agency and the European Commission was also very helpful, and the ECMT thanks collaborators in these organisations who provided insight and comments to the reports.

Mary Crass, environmental policy analyst and consultant with the ECMT prepared the first report, "Monitoring of National Policies for the Reduction of CO_2 Emissions" under the supervision of the ECMT group on transport and the environment. The second report was prepared by the ECMT joint industry-government group and the ECMT is particularly grateful for the support of John Turner on behalf of the European Automobile Manufacturers Association (ACEA) and Emilio di Camillo from the International Organization of Motor Vehicle Manufacturers (OICA).

FOREWORD

The transport sector is a principal source of greenhouse gas emissions -- in particular CO_2 -- and is therefore a main contributor to potential climate change. Given its key role in the problem, transport has been a centre of focus in the policy debate surrounding climate change. What to do about transport-related emissions has been difficult for policy makers to come to terms with. Rapidly rising demand for mobility around the world -- especially for road transport -- has rendered strong policy action difficult to take. It is also complicating the task of meeting emissions stabilisation objectives established at the Earth Summit in Rio de Janeiro in 1992.

Transport Ministers have recognised the significant role that the transport sector plays in climate change, and have taken steps within the European Conference of Ministers of Transport (ECMT) to explore ways to address the issue. ECMT countries submit their plans to limit CO_2 emissions from transport for periodic review and Ministers have established an on-going dialogue with motor vehicle manufacturers and importers to examine ways to limit CO_2 emissions from new cars. A joint declaration on reduction of CO_2 was signed by Ministers and industry representatives in 1995.

In response to Ministers requests, the ECMT conducted an in-depth survey in 1996 of its Members and Associate Members to find out how countries were responding to the challenge of reducing transport-related CO_2. ECMT countries were asked to provide information on:

-- CO_2 emissions data from the transport sector disaggregated to subsector level; and
-- transport policy actions either in effect or planned to limit CO_2 emissions.

The first part of this publication, the report "Monitoring of National Policies for the Reduction of CO_2 Emissions from Transport", contains the conclusions of this survey. Among other findings, the survey showed that despite policy initiatives in many ECMT countries to limit CO_2 emissions from transport, these emissions will continue to rise in both relative and absolute terms up to 2010. And commitments already made in the context of the Framework Convention on Climate Change will only in very few cases be met.

The dialogue with industry has centred on methodologies for tracking fuel consumption and CO_2 emissions from new cars. The second part of this publication, the report entitled "Monitoring of Fuel Consumption and CO_2 Emissions of New Cars", examines the requirements of a monitoring system and takes a look at current data sources. It concludes that while data remain imperfect, they are sufficient to record trends in new car fuel consumption to the degree of accuracy required.

At the 1997 Ministerial session in Berlin, Ministers noted both of these reports, and agreed to their recommendations and proposed follow-up actions.

The ECMT has co-operated closely with other international bodies in seeking policy solutions to transport-related CO_2 emissions, and will continue to participate in the on-going climate change debate. The ECMT hopes that this publication will further understanding of transport's role in climate change and stimulate ideas that will move the policy debate forward.

TABLE OF CONTENTS

EXECUTIVE SUMMARY 9

MONITORING OF NATIONAL POLICIES FOR REDUCTION
OF CO_2 EMISSIONS 13

    1. INTRODUCTION 19
    2. CO_2 EMISSIONS WORLD-WIDE 24
    3. CO_2 EMISSIONS IN ECMT COUNTRIES 27
    4. POLICIES AND MEASURES TO LIMIT CO_2 FROM TRANSPORT 53
    5. CONCLUSIONS 116
    NOTES 119
    ANNEXES... 121
    BIBLIOGRAPHY 167

MONITORING OF FUEL CONSUMPTION AND CO_2 EMISSIONS
OF NEW CARS 173

    A. BACKGROUND 177
    B. TRENDS 179
    C. MONITORING METHODOLOGY 187
    D. DATA 195
    NOTES 202
    ANNEX: Joint declaration on reducing carbon dioxide emissions
                   from passenger vehicles in ECMT countries... 203

EXECUTIVE SUMMARY

The Role of Transport in Climate Change

The transport sector's role in climate change presents a particular challenge to policy makers. Transport—notably road transport—is responsible for significant levels of anthropogenic greenhouse gases: carbon dioxide (CO_2), methane, chlorofluorocarbons and low-level ozone precursors such as hydrocarbons and nitrous oxides. Emissions of these gases occur during the production and processing of fossil fuels; in their combustion as a part of the operation of transport systems; and in the manufacture and disposal of transportation equipment. Road vehicles, trains, ships and aircraft account for approximately 30 per cent of CO_2 emissions from fossil fuel combustion in the OECD (including refinery and power station emissions from energy production for the transport sector).

Transport's share in the climate change problem is especially difficult to address because of rapidly rising demand for mobility -- particularly for road transport -- which is the principal contributor to potential transport-related global warming. Road transport accounts for approximately 80 per cent of greenhouse gases from transport in Western Europe; of this, two-thirds is attributable to the car -- primarily in the form of CO_2. Travel by car has increased substantially over the last 50 years in OECD and ECMT countries, as incomes have grown and lifestyles have become increasingly car-dependent.

The International Climate Change Context

These current trends in transport emissions appear to be incompatible with the emissions stabilisation objectives established at the United Nations Conference on the Environment and Development (UNCED) in Rio de Janeiro in June 1992. The Framework Convention on Climate Change (FCCC) was signed by 154 countries and the European Economic Community (now the European Union) at UNCED. As of March 1997, 164 parties had signed and ratified the Convention. For industrialised countries, the initial goal of the convention was to stabilize emissions of CO_2 and other greenhouse gases at 1990 levels by 2000. The great majority of ECMT Member and Associate Member countries have signed and ratified the FCCC, and are thus held to its requirements to provide information on greenhouse gas emissions inventories and climate change policies.

At the first Conference of Parties to the Convention (COP-1), which took place in Berlin in 1995, it was concluded in the "Berlin Mandate" that current commitments of developed countries to take measures aimed at meeting FCCC targets were inadequate and that talks on a new protocol or legal instrument providing for stronger commitments after the year 2000 were necessary. In July 1996, the Conference of Parties met for a second time (COP-2) in Geneva, where a declaration calling for the setting of quantified, legally-binding targets on greenhouse gas emissions beyond 2000 was drawn up. At the time of publication, these commitments were the subject of on-going negotiations and were scheduled to be adopted at the third Conference of Parties in Kyoto, Japan in December 1997.

Principal Findings of ECMT Work

1. Monitoring of National Policies for the Reduction of CO_2 Emissions from Transport

Under mandate from the 1994 Council of Ministers session in Annecy, the ECMT Secretariat consulted Member and Associate Member countries and the European Union in 1996, requesting updated information on CO_2 emissions from transport and on policies and measures in place or envisaged to limit these emissions. Twenty-nine of the 38 countries responded to the survey. Based on the responses, the following can be concluded:

- A small number of countries have developed focused strategies for CO_2 abatement from transport. The majority of countries have not yet put such strategies in place.
- Many ECMT countries have implemented or are considering a variety of measures that have been shown to be effective in CO_2 abatement, notably carbon or CO_2 taxes, voluntary agreements on fuel efficiency with vehicle manufacturers, higher fuel taxes, tighter speed limit enforcement, and actions to improve driver behaviour.
- In spite of these efforts, the data show that transport sector emissions will continue to rise both in relative and absolute terms through 2010 in almost all ECMT countries, and that, largely as a consequence, overall greenhouse gas stabilisation targets will most likely not be met by a large majority of ECMT members.
- In their responses to the survey, countries listed numerous policies and measures as contributing to CO_2 reduction, including general policy principals such as "promoting railway use", "supporting public transport" and "enforcing parking restrictions". In most cases, the effectiveness of such general policies for reducing CO_2 emissions has yet to be demonstrated. Much remains to be understood as to how different policies work individually and in combination to achieve stabilisation and reduction of CO_2 emissions.
- Countries are working to render their databases more disaggregated, reliable, and sophisticated in terms of the information they produce. Many data and methodological problems remain, however, and until improvements in these areas are made, comparative analyses will be difficult.

2. The Dialogue with Industry

At the session in Vienna in 1995 Ministers and the Vehicle Manufacturing Industry (OICA and ACEA) adopted a joint agreement with the objectives of "substantially and continuously reducing the fuel consumption of new passenger cars sold in ECMT Countries" and "managing vehicle use so as to achieve tangible and steady reductions in their CO_2 emissions". Since 1995, a number of activities have been carried out in support of the agreement.

- A first report on monitoring has been prepared and is included in this publication. The report sets out the requirements of a monitoring system and examines the adequacy of present data sources. An existing international industry data base is being used to provide indications of trends in fuel consumption. While not perfect (for example, not all countries are included and not all sources use the same definitions) the data are sufficiently accurate to provide an adequate picture of the trends in new car fuel consumption. This is a long-term issue and not enough time has elapsed to determine trends in specific fuel consumption since adoption of the joint agreement.

- A workshop on improving driver behaviour through use of in-car equipment (econometers, cruise controls, on-board computers) and driver training was held in Delft at the invitation of the Government of the Netherlands. The findings show that there is valuable potential for reducing fuel consumption by improving driver behaviour, and indicate how this might be achieved through a combination of measures.
- An examination of car advertising was also carried out. The study revealed that a marked change has taken place in the advertising of cars in recent years; notably, that power and speed characteristics are emphasised less explicitly than in the past, partly due to actions by advertising standards bodies and voluntary agreements with industry (originating in safety concerns). While there could still be improvements, it was agreed that it would not be worthwhile to try and draw up a code of practice at international level. The significant influence of motoring journalists in shaping consumer demand was also identified.

As part of its CO_2 strategy, the EU Commission has begun discussions with the vehicle manufacturing industry (ACEA) with a view to negotiating an agreement on CO_2 emissions from cars. In this context, technical work is under way with EU Member states on a comprehensive official monitoring system. ECMT has been co-operating with the Commission in the development of monitoring methodologies.

Main Recommendations

- Countries should make efforts to develop a more strategic approach to CO_2 abatement. This requires designing cost-effective packages of policies and measures to reduce CO_2 emissions from transport as an integrated part of economy-wide measures to reduce greenhouse gas emissions. It also requires balancing CO_2 emissions reduction with other key transport policy goals. The currently fragmented public support for Research and Development programmes also needs to be more strategically focused.
- Countries should work towards developing a better understanding of the links between specific policy measures and CO_2 emissions, with a view to adopting measures that are effective in limiting CO_2 emissions.
- In the short-term, countries should seek and seize opportunities to implement cost-effective or "win-win" measures that can be introduced early on and which have benefits in other spheres such as local air pollution abatement, road safety and efficiency in the transport sector, as well as limiting CO_2 emissions. Examples of such measures might include: stricter enforcement of speed limits; tighter vehicle inspection systems; information campaigns and education to improve driver behaviour; efficient structures of fuel and vehicle taxation; better fleet management and improved vehicle loading factors.
- In addition, countries should take further action to improve their data collection and monitoring of CO_2 emissions from the transport sector and its component parts.

Decisions of Transport Ministers Regarding CO_2 Emissions at the 1997 Ministerial Session in Berlin

Ministers in Berlin agreed to the recommendations outlined in the above section and to the proposals described below:

ECMT should:

- Submit these conclusions, and the full reports on which they are based, to the Framework Convention on Climate Change and the meeting of the Conference of the Parties to the Convention in Kyoto in December 1997, as well as to the November 1997 UN/ECE Conference in Vienna, as a contribution from Ministers of Transport.

- Continue to exchange information and experience and conduct targeted analysis on national and international policies and measures and their effectiveness in limiting CO_2 emissions. Report on these items as appropriate to the Council of Ministers.

- Repeat the survey of national policies for reducing CO_2 emissions in transport after a period of about four years.

- Continue the dialogue with industry, in close co-operation with the European Union and other organisations. Work under the dialogue should focus on:
  - appropriate incentives for affecting consumer choice and overcoming barriers to the introduction of more fuel efficient vehicles;
  - reducing CO_2 emissions from freight vehicles;
  - the role of telematics and infrastructure improvements in reducing CO_2 emissions;
  - providing a progress report in 1999.

MONITORING OF NATIONAL POLICIES FOR REDUCTION OF CO_2 EMISSIONS

TABLE OF CONTENTS

1. INTRODUCTION 19
   1.1   The role of transport in climate change 19
   1.2   The ECMT mandate 20
   1.3   The Framework Convention on Climate Change (FCCC) context 20
   1.4   Methodology of the 1996 survey 22
   1.5   Structure of the Report 23

2. CO_2 EMISSIONS WORLD-WIDE 24
   2.1 CO_2 emissions within the OECD 25

3. CO_2 EMISSIONS IN ECMT COUNTRIES 27
   3.1   Brief country profiles 27
   3.2   Summary of emissions data 46

4. POLICIES AND MEASURES TO LIMIT CO_2 FROM TRANSPORT 53
   4.1   Description of policies and measures targeting CO_2 from transport 53
   4.2   Policy initiatives in ECMT countries 65
   4.3   Summary and remarks 107

5. CONCLUSIONS 116

NOTES 119

ANNEXES 121
1. Questionnaire 123
2. The Framework Convention on Climate Change: Status of ECMT Countries 133
3. Comparative data tables 134

BIBLIOGRAPHY 167

LIST OF TABLES IN TEXT

CO_2 Emissions Data

1. CO_2 emissions world-wide 25

2. CO_2 emissions from fossil fuel combustion in OECD countries 26

3. Summary of CO_2 emissions in ECMT countries 47

Policies and Measures to Limit CO_2 from Transport

4a to 4g   Economic Instruments 67

5a to 5f   Regulations and Guidelines 86

6a to 6c   Voluntary Agreements/Actions 98

7a to 7d   Information and Training Initiatives 100

8a to 8d   Research and Development 103

9.   Responses to Table 7 of the Questionnaire on Policies and Measures 108

10.   Share of Types of Policies and Measures in Total 109

11a to 11e   Policy Categories by Country 110

LIST OF TABLES IN ANNEX

12. Base-year Emissions of CO_2 by Sector 135

13. Base-year Emissions of CO_2 by Sub-sector 136

14. Reference Case: Projected Emissions of CO_2 2000, 2010 by Sector 139

15. Reference Case: Projected Emissions of CO_2 2000, 2010 by Sub-sector (Road) 140

16. Reference Case: Projected Emissions of CO_2 2000, 2010 by Sub-sector (Shipping, Aviation) 141

17. Status Quo Case: Projected Emissions of CO_2 2000, 2010 by Sector 144

18. Status Quo Case: Projected Emissions of CO_2 2000, 2010 by Sub-sector (Road) 145

19. Status Quo Case: Projected Emissions of CO_2 2000, 2010 by Sub-sector (Shipping, Aviation) 146

20. Future Measures Case: Projected Emissions of CO_2 2000, 2010 by Sector 148

21. Future Measures Case: Projected Emissions of CO_2 2000, 2010 by Sub-sector (Road) 149

22. Future Measures Case: Projected Emissions of CO_2 2000, 2010 by Sub-sector (Shipping, Aviation) 150

23. Projected Emissions of CO_2 2000, 2010 from Road Transport by Fuel 153

24. Annual CO_2 Emissions Data from Road Transport 1990-1995 155

25. Annual CO_2 Emissions Data from Rail 1990-1995 157

26. Annual CO_2 Emissions Data from Shipping 1990-1995 158

27. Annual CO_2 Emissions Data from Aviation 1990-1995 159

28. Annual CO_2 Emissions Data from Other Categories 1990-1995 160

29. Annual CO_2 Emissions: Transport Sector and Total 1990-1995 161

30. Annual CO_2 Emissions Data by Fuel 1990-1995 164

1. INTRODUCTION

1.1 The role of transport in climate change

The transport sector's role in climate change is particularly challenging to policy makers. Transport -- notably road transport -- is responsible for significant levels of anthropogenic greenhouse gases: carbon dioxide (CO_2), methane (CH_4), chlorofluorocarbons (CFC) and low-level ozone precursors such as hydrocarbons (HC) and nitrous oxides (N_2O). Emissions of these gases occur during the production and processing of fossil fuels; in their combustion as a part of the operation of transport systems; and in the manufacture and disposal of transportation equipment. Road vehicles, trains, ships and aircraft account for approximately 30 per cent of carbon dioxide emissions from fossil fuel combustion in the OECD (including refinery and power station emissions from energy production for the transport sector). Current trends in emissions from transport appear to be incompatible with the emissions stabilization objectives established at Rio de Janeiro in 1992[1].

The significant role of transport in the climate change problem is exacerbated by rapidly rising demand for mobility -- especially for road transport, which is the principal contributor to potential transport-related global warming. Road transport accounts for approximately 80 per cent of greenhouse gases from transport in Western Europe; of this, two-thirds is attributable to the car -- primarily in the form of CO_2. Travel by car has increased substantially over the last 50 years in OECD countries, as incomes have grown and lifestyles have become increasingly car-dependent. Growth in car use varies in different regions of the OECD: in the United States, for example, per capita vehicular travel is twice that of European countries and transport energy demand three times higher. In Japan, vehicular travel per capita is lower than in Europe, despite higher output per head.

Road freight traffic has seen a nearly 5 per cent annual growth rate over the last 20 years, surpassing car traffic in the same time period (3.3 per cent per year). Air transport of both passengers and goods is also on the rise -- a particular concern as regards climate change, because air travel is a significant emitter of high-level ozone precursors. Most industrialised countries expect the transport sector to be the most rapidly growing source of greenhouse gas emissions up to the year 2000, and the International Panel on Climate Change's Second Assessment Report identifies transport as the fastest growing sector to 2025 (ECMT, 1993a).

At present, the majority of the world's car fleet and transport-related environmental problems are found in developed countries. However, the trend towards higher demand for cars is rapidly spreading to countries with transitional economies in Central and Eastern Europe and in the developing world.

Faced with both high demand for road transport and growing concern over climate change, government and industry decision makers around the world are seeking policy options to address these problems. International bodies including the European Conference of Ministers of Transport

(ECMT) are actively engaged in the dialogue on policies and measures to address the problem of transport-related CO_2 emissions.

1.2 The ECMT mandate

The ECMT has been increasingly active in the debate on global warming. In a 1989 ECMT Resolution on Transport and the Environment, Ministers set as a priority that "a full range of possible measures that can be taken to reduce transport's contribution to the 'greenhouse effect' be set out together with the costs and practical problems of implementing them" (ECMT, 1989). A subsequent 1991 resolution recommends that the need for regulations on maximum vehicle power-to-weight ratios be the focus of UN/ECE and EC studies and that taxation of vehicles and fuels should be "consistent with and reinforce the policy aim of limiting the growth in vehicle power and potential speed" (ECMT, 1991).

In response to these resolutions, an International Seminar on Reducing Transport's Contribution to Global Warming was held in 1992, which led to further commitment from the ECMT to address the climate change issue: ECMT Member countries were requested in 1993 to provide information on their programmes to stabilize and reduce transport-related carbon dioxide emissions (ECMT, 1993b). An interim report on these initiatives was prepared with a view to a follow-up survey two years later.

In May and June 1996, the ECMT sent a questionnaire to its Member and Associate Member countries[2] requesting updated information on their CO_2 emissions from transport and on policies and measures in place or envisaged to limit CO_2 from transport (see Annex 1). As of February 1997, 29 of the 38 recipient countries of the questionnaire had responded.

The findings of this report are based on the information provided in the responses to the Spring 1996 questionnaire.

1.3 The Framework Convention on Climate Change (FCCC) context

In carrying out the present survey on CO_2 emissions and policies, it was important to consider initiatives already under way within the context of the FCCC. The vast majority of ECMT Member and Associate Member countries have signed and ratified the FCCC (See Annex 2), and are thus held to the convention's specific requirements regarding the provision of information on greenhouse gas emissions inventories and policies and measures targeting these emissions. In order to set the ECMT work in perspective relative to this larger FCCC context, a brief description of the FCCC process, recent developments and future expectations follows.

1.3.1 The FCCC process

In June 1992, 154 countries and the European Economic Community (now the European Union) signed the FCCC at the United Nations Conference on the Environment and Development in Rio de Janeiro, Brazil. The FCCC entered into force 90 days after its ratification by the fiftieth signatory of the Convention on 21 March 1994, and thus became international law for those parties. As of March 1997, 164 parties had ratified the Convention.

The overall objective of the FCCC is the "stabilization of greenhouse gas concentrations in the atmosphere at a level that would prevent dangerous anthropogenic interference with the climate system" (IEA/OECD 1994). For industrialised countries (those often referred to as "Annex 1" countries), the initial goal set forth in the FCCC was to stabilize emissions of CO_2 and other greenhouse gases at 1990 levels by 2000. Five years later, it appears clear that very few of these countries will actually meet that target.

Moreover, at the first Conference of Parties (COP-1), which took place in Berlin in March-April 1995, it was concluded in what is now known as the "Berlin Mandate" that the current commitment of developed countries to take measures aimed at meeting this target was inadequate, and that talks on a new protocol or legal instrument providing for stronger commitments from these countries after the year 2000 were necessary. These new commitments are the subject of current negotiations within the Ad-hoc Group on the Berlin Mandate (AGBM), which was established at COP-1 as the forum for debate on stronger commitments from industrialised countries. The AGBM process is to lead developed countries to "elaborate policies and measures" and "set quantified limitation and reduction objectives within specified time-frames, such as 2005, 2010 and 2020 for their anthropogenic emissions". The new commitments are to be adopted at the third Conference of Parties, which will take place in Kyoto, Japan in December 1997. The Berlin Mandate does not call for any new commitments from developing and newly industrialising countries (so-called non-Annex 1 countries), but calls on them to "advance the implementation" of their existing commitments.

1.3.2 *Recent key developments and outlook*

In July 1996, the Conference of Parties met for a second time (COP-2) in Geneva, marking the mid-point in the schedule for meeting the objectives of the Berlin Mandate. Politically significant were the formulation of the Geneva Declaration, which mandates the setting of quantified, legally-binding targets on greenhouse gas emissions beyond 2000[3], and the shift in the position of the United States in favour of a "realistic, binding target" for greenhouse gas emissions[4]. In addition, the much-contested Second Assessment Report of the International Panel on Climate Change was adopted, which concludes, *inter alia*, that human activity -- primarily fossil fuel combustion -- will interfere with the global climate system (UN-FCCC, 1996).

It is clear that despite the fact that most countries will not meet their current commitments to the FCCC, new steps are being taken to reduce the risk of climate change. Governments are studying a variety of policy approaches, notably as regards transport, to deal with CO_2 and other greenhouse gas emissions: many have already taken politically bold regulatory and fiscal steps to meet their commitments; others are taking a slower, more hesitant approach. It is in this climate, that the second ECMT survey on CO_2 emissions from transport was carried out.

1.3.3 *FCCC link with the 1996 ECMT survey*

Given the complementary nature of the information requested from countries in the ECMT survey and that required by the FCCC, efforts were made to render the ECMT questionnaire generally compatible with the information required under the FCCC. In the policies and measures section of the questionnaire, for example, initial information provided to countries for confirmation and updating was largely drawn from the national communications to the FCCC. Further, throughout the preparation of this report, the ECMT secretariat exchanged information with both the FCCC secretariat and OECD and International Energy Agency experts supporting the Annex 1 Expert Group, an ad-hoc group established in 1993 by the OECD Member countries. ECMT also presented a paper on policies and measures to limit CO_2 from transport at a meeting in Geneva of the AGBM in March 1996 and to a national climate change conference in Canada in September 1996. By working the ECMT report into the larger FCCC initiative, it was hoped that not only both processes would be benefited, but also that the time and effort required of the countries to respond to the ECMT questionnaire would be minimised.

1.4 Methodology of the 1996 survey

The questionnaire (see Annex 1) sent to ECMT Members and Associate members and the European Union solicited two principal kinds of information: the first, CO_2 emissions inventories for the transport sector; the second, policies and measures in place or envisaged to limit CO_2 emissions from transport. The structure of the questionnaire is as follows:

a. Information requested on emissions

1.1 Base-year data
1.2 Projected CO_2 emissions data for 2000, 2010: Reference Case
1.3 Projected CO_2 emissions data for 2000, 2010: Status Quo Case
1.4 Projected CO_2 emissions data for 2000, 2010: Future Measures Case
1.5 CO_2 emissions data from the Road Transport sub-sector by fuel
1.6 Annual emissions data from sector and by fuel

b. Information requested on policies and measures

For all policies and measures using defined abbreviations:

Objective, Instrument/Approach, Status, Quantitative Analysis, Cross-References to data tables.

Although compatibility with the requirements of the FCCC was a concern in the development of the questionnaire, it diverges from the FCCC approach in a number of ways, notably:

-- it goes beyond both the FCCC requirements and the ECMT survey carried out in 1993 in terms of the degree of disaggregation requested regarding both transport-related CO_2 emissions and information concerning the impact of policies and measures on emissions levels;
-- in addition to Reference and Future emissions scenarios, it requests projections for the years 2000 and 2010 for both sub-sectors and fuels and includes a Status Quo case, which are not required under the FCCC;
-- international bunker emissions are included in overall emissions totals, whereas they are not in the FCCC context.

1.4.1   The data in perspective

The objective in requesting more detailed information than that required by the FCCC was to try to obtain as comprehensive a view as possible of:

-- where ECMT countries stand at present regarding CO_2 from transport; and
-- what kinds of developments have occurred since the preparation of the interim report in 1993.

After examination of the responses received, it can be concluded that the totality of information solicited in the questionnaire was available only in very few cases. Minimum requirements for the FCCC appear to be guiding emissions inventory development. Efforts to improve information collection within the countries seem to be taking place, however, both as regards the level of sectoral disaggregation of emissions inventories, and the analysis of the impact of policies and measures on CO_2 emissions.

Lack of transparency in data and incomplete information in general render analysis and comparisons among countries difficult, particularly as concerns the:

-- methodology used to obtain emissions data;
-- nature of policies and measures to limit CO_2 emissions; and
-- way measures are/will be used, as well as their real or projected impacts.

The FCCC has brought to light this problem as well in their first review of the national communications on climate change. The FCCC notes that many of the Annex 1 countries that have completed their communications have not followed the minimum documentation standards or provided explanations of methods and data according to the FCCC guidelines. Moreover, the variety of assumptions and definitions among different countries impedes analysis on policies and emissions (UN-INC, 1995). Clearly, improvement of the methodologies is needed, as well as the adherence to these methodologies.

Regarding the reliability of data, countries which have submitted communications to the FCCC have generally indicated a high confidence level in CO_2 data relative to data on other greenhouse gases inventoried (UN-INC, 1995).

1.5 Structure of the report

The structure of this report is as follows: in Section 2, emissions of CO_2 will be briefly examined, first from a global and regional perspective to set the ECMT situation in context, then in Section 3 on an ECMT level based on the responses to the questionnaires. Section 4 will explore policies and measures to limit CO_2 in terms of available policy options and how ECMT countries are using these options to address CO_2 emissions from transport.

2. CO_2 EMISSIONS WORLD-WIDE

In spite of current efforts to stabilize CO_2 emissions, global levels of CO_2 are expected to significantly rise by the year 2010 relative to 1990. As shown in Table 1, the International Energy Agency (IEA) projects an increase of 50 per cent in CO_2 during this period under the IEA's Capacity Constraint scenario (see Explanatory Note below) and by roughly 36 per cent in the Energy Savings case. According to IEA projections, the majority of the growth in emissions will take place in countries outside of the OECD area and Central and Eastern Europe – the «Rest of the World» -- where emissions are expected to more than double under each scenario.

Explanatory note for Table 1 :
The IEA uses two scenarios for projections of growth in CO_2 emissions: *Capacity Constraints and Energy Savings:*

-- *Capacity Constraints* assumes combined GDP and population assumptions with rising energy prices and historical trends in energy efficiency. A moderation in energy demand to 2010 is assumed to take place though an increase in primary energy prices (or constant primary energy prices and increasing energy taxation). The idea is that available capacity is insufficient to meet growth in demand despite adequate resource supply (as has been the case over the last two decades). The result is an increase in prices and a moderation in energy demand.

-- *Energy Savings* assumes an increase in energy efficiency combined with baseline economic growth and population assumptions and flat energy prices. The greater efficiency in energy use is not assumed to derive from new technologies, but rather from energy consumption choices made by industrial and household consumers that reflect improved use of best commercially available technologies and more-efficient use of energy equipment. Potential energy savings from this improved consumer behaviour is assumed to be very large (IEA/OECD 1996a).

Table 1. CO_2 emissions world-wide
(billion tonnes of CO_2)

	1990	2000		2010	
		Energy Savings	Capacity Constraints	Energy Savings	Capacity Constraints
OECD	**10.4**	**11.3**	**11.8**	**11.8**	**13.3**
North America	5.6	6.2	6.5	6.6	7.3
Europe	3.4	3.5	3.6	3.6	4.0
Pacific	1.4	1.6	1.7	1.7	1.9
Former Soviet Union / Central & Eastern Europe	**4.4**	**3.2**	**3.4**	**3.6**	**4.2**
Rest of the World	**5.9**	**8.8**	**9.0**	**12.8**	**13.5**
China	2.4	3.4	3.5	5.0	5.1
East Asia	0.9	1.6	1.7	2.4	2.5
South Asia	0.7	1.1	1.1	1.9	2.0
Other	1.9	2.7	2.7	3.5	3.9
World	**21.1**	**23.7**	**24.7**	**28.8**	**31.5**

Notes:
1. Totals have been rounded to nearest tenth.
2. World totals include bunkers.

Source: IEA/OECD, 1996a.

2.1 CO_2 emissions within the OECD

Within the OECD, North America is likely to account for the largest share of the increase in CO_2 emissions in both the Capacity Constraints and Energy Savings scenarios, as the bulk of energy in the OECD is consumed in this region. Growth in CO_2 is expected to be fastest in percentage terms, however, in the Pacific region of the OECD, largely due to growth in Japan's energy demand relative to a decreasing share of overall energy demand in OECD North America and Europe (IEA/OECD 1996a).

Table 2 shows the comparison between CO_2 emissions in OECD countries from 1990, the base year for most countries, and 1994.

Table 2. CO_2 emissions from fossil fuel combustion in OECD countries[1]

	Total (Mt CO_2)		(Kt CO_2) per capita		(Kt CO_2) per GDP (1000 US $ 1990)		Transport					
							Total Transport (Mt CO_2)		Per capita Transport (Kt CO_2)		Per cent Transport	
	1990	1994	1990	1994	1990	1994	1990	1994	1990	1994	1990	1994
Australia	265.38	292.07	15.6	16.4	0.90	0.90	66.04[3]	69.84	3.9	3.9	24.9	23.9
Austria	59.39	57.81	7.7	7.2	0.37	0.34	15.71[3]	17.69[3]	2.0	2.2	26.5	30.6
Belgium	109.47	117.34	11.0	11.6	0.57	0.58	23.16	25.55	2.3	2.5	21.2	21.7
Canada	431.71	460.14	16.2	15.7	0.76	0.76	126.95[2]	134.40[2]	4.8	4.6	29.4	29.2
Denmark	53.37	64.31	10.4	12.4	0.41	0.46	13.58[3]	13.83	2.6	2.7	25.4	21.5
Finland	53.75	61.16	10.8	12.0	0.40	0.49	12.81	12.48	2.6	2.5	23.8	20.4
France	379.28	347.97	6.7	6.0	0.32	0.28	124.90	134.94	2.2	2.3	32.9	38.8
Germany	983.10	888.04	12.4	10.9	0.60	0.51	173.50	183.35[3]	2.2	2.3	17.6	20.6
Greece	72.61	77.40	7.2	7.4	0.89	0.90	17.62	19.51	1.7	1.9	24.3	25.2
Iceland	2.43	2.46	9.5	9.2	0.39	0.38	857.31 (Kt)	875.92 (Kt)	3.3	3.3	35.3	35.6
Ireland	33.32	34.69	9.5	9.7	0.74	0.66	6.00	6.97	1.7	2.0	18.0	20.1
Italy	408.18	400.92	7.2	7.0	0.37	0.35	99.81	109.82	1.8	1.9	24.4	27.3
Japan	1067.92	1 142.05	8.6	9.1	0.36	0.36	213.95[3]	243.27	1.7	1.9	20.0	21.3
Luxembourg	10.86	10.95	28.4	27.5	1.05	0.98	3.05	4.06	8.1	10.2	28.0	37.0
Mexico	308.24	353.79	3.8	4.0	1.26	1.31	94.62	105.25	1.2	1.2	30.7	29.7
Netherlands	161.50	174.67	10.8	11.4	0.57	0.57	30.73	35.18	2.1	2.3	19.0	20.1
New Zealand	25.34	27.85	7.5	7.9	0.58	0.56	10.38[3]	11.87[3]	3.1	3.4	40.9	42.6
Norway	31.70	33.22	7.5	7.7	0.27	0.25	12.31	13.77	2.9	3.2	38.8	41.4
Portugal	41.58	46.57	4.2	4.7	0.62	0.67	11.30	14.15	1.1	1.4	27.1	30.4
Spain	217.42	237.16	5.6	6.1	0.44	0.46	67.15	77.10	1.7	2.0	30.9	32.5
Sweden	52.81	55.17	6.2	6.3	0.23	0.25	21.13[3]	22.08	2.5	2.5	40.0	40.0
Switzerland	44.36	43.84	6.6	6.3	0.20	0.19	17.73	18.31	2.6	2.6	39.9	41.8
Turkey	138.48	146.67	2.5	2.4	0.92	0.89	28.40	32.36	0.5	0.5	20.5	22.0
United Kingdom	584.02	563.38	10.2	9.7	0.60	0.56	135.40	139.90	2.4	2.4	23.1	25.0
United States	4894.54	5 160.1	19.6	19.8	0.89	0.86	1455.45	1537.93	5.8	5.9	29.7	30.0
OECD Total	10 430.76	10 799.73	9.8	9.9	0.59	0.60	2 782.54	2 984.48	2.67	2.86	27.7	29.1

Sources: Emissions figures derived from IEA annual data on CO_2 emissions from fossil fuel combustion, August 1996.
GDP and per capita figures derived from OECD data.

Notes:
1. Data represent emissions from fuel combustion in road, rail, internal navigation, domestic air transport, international civil aviation.
2. Includes pipeline transport.
3. Includes other non-specified transport.
4. The divergences in the numbers presented here and those submitted by ECMT countries in their responses to the questionnaire may be explainable for several reasons, among them: the questionnaire explicitly requested that international bunker fuels be included in totals; the IEA includes aviation bunker fuels, but excludes marine bunker fuels from national energy consumption and emissions. Secondly, according to the International Energy Agency, the IPCC/IEA default methodology uses a default carbon emissions factor that neither accounts for variability in the carbon content of coal and other fuels, nor considers differences in refinery specifications from country to country, which leads to different levels of combustion.

3. CO_2 EMISSIONS IN ECMT COUNTRIES

This section is based primarily on the information provided in responses to the 1996 survey. Of the 38 recipients of the questionnaire, 29 had responded by February 1997.

Questionnaires were returned in complete or partial form by:

Austria, Belgium, Canada, Czech Republic, Denmark, Finland, France, Germany, Hungary, Ireland, Italy, Japan, Latvia, Lithuania, the Netherlands, New Zealand, Norway, Poland, Portugal, Romania, the Russian Federation, Slovak Republic, Slovenia, Spain, Sweden, Switzerland, United Kingdom, United States and the European Union.

Correspondence indicating the limited nature of the information available to respond to the questionnaire was received from:

Croatia and Turkey.

Responses were not received from:

Australia, Bosnia-Herzegovina, Bulgaria, Estonia, Greece, Luxembourg and Moldova.

3.1 Brief country profiles

This section provides basic information on climate change targets, GDP and population data and basic trends in the transport sector for each of the country respondents to the questionnaire. The information is not meant to be comprehensive; the profiles are designed to simply lend context to the CO_2 emissions data and the descriptions of policies and measures that follow in the rest of the report.

Austria		
	Approach to Climate Change	**Target**: 20% reduction in CO_2 emissions by 2005 relative to 1988 (Toronto target) • In November 1992, the Government accepted the Toronto target, which applies to all anthropogenic CO_2 emissions, but particularly emphasizes energy-related emissions. • Austria signed the UN FCCC at the Rio Conference in 1992 and ratified it on 28 February 1994.
	GDP 1994 (billion US$, 1990)	171.9
	Population 1994 (million)	8.03
	Principal Characteristics and Trends in the Transport Sector	⇒ <u>Freight market share</u>: 41% rail; 31% road; 5% inland navigation, 23% pipeline ⇒ <u>Passenger market share</u>: 12% rail; 8% bus/coach ⇒ <u>No. of private</u> vehicles: 260 000 in 1950; 4.2 million in 1990 ⇒ <u>Road traffic volume</u> has increased rapidly over last 20 years: 80% increase in v-km travelled 1980-1991 (Austria, 1994 and OECD, 1995b).

Belgium		
	Approach to Climate Change	**Target**: 5% reduction in total anthropogenic CO_2 emissions from 1990 levels by 2000. This target was adopted by the Council of Ministers in June 1991. • Belgium signed the FCCC in Rio in 1992 and ratified the Convention in January 1996.
	GDP 1994 (billion US$, 1990)	201.0
	Population 1994 (million)	10.1
	Principal Characteristics and Trends in the Transport Sector	⇒ <u>Highway traffic</u>: 68.8% increase; 37.8% on other roads from 1985 to 1994. ⇒ <u>Heavy vehicle traffic</u> accounted for 15.3% of highway traffic in 1993, up from 14.75% in 1990; this increase was greater than that of light vehicle traffic. ⇒ <u>Rail passenger traffic</u> increased only slightly from 1990 to 1994, whereas rail *freight* transport dropped 10% in 1993 relative to 1991-92. ⇒ <u>No. of private vehicles</u>: In August 1995, 466 vehicles/1 000 habitants, 421 of which, cars (Belgium, 1996).

Canada		
	Approach to Climate Change	**Target**: Stabilization of the net[*] aggregate global warming potential of CO_2 and other non-Montreal Protocol greenhouse gases by 2000 at 1990 levels. This target was established in Canada's *Green Plan* of 1990. • Canada signed the FCCC in Rio in 1992 and ratified it in December of the same year.
	GDP 1994 (billion US$, 1990)	601.4
	Population 1994 (million)	29.2

---

* "Net" means the carbon removal by sinks is subtracted from direct emissions.

Canada cont'd	Principal Characteristics and Trends in the Transport Sector	⇒ In 1995, <u>passenger transport</u> totalled 500 billion passenger-km or 17 000 passenger-km per capita. <u>Private vehicles</u> accounted for 94% of the total, and <u>public transport</u> only 6%. <u>Travel in urban areas/rural municipalities</u> totalled nearly 300 billion passenger-km, with urban transit accounting for 5 billion -- less than 2%. Total intercity traffic totalled approximately 200 billion passenger-km, private vehicles again accounting for a predominant 87%, air travel -- 11% and intercity bus and train -- 2%. ⇒ As concerns <u>domestic freight</u>, 1995 traffic amounted to approximately 410 billion tonne-km. <u>Rail traffic</u> dominated domestic freight transport, with about 50% of total traffic, and trucking carried around 40% (based on preliminary private trucking figures). Marine transport accounted for approximately 10% of domestic freight carriage. ⇒ Forecasts for average annual growth in <u>freight transport</u> 1996 to 2005 are as follows: for-hire trucking 2.3%; rail 1.2% and marine traffic 1.1%. ⇒ As concerns <u>passenger transport</u>, <u>air traffic</u> to the United States is projected to grow 4.9% annually from 1996 to 2005; 5.1% to other international destinations and 2.3% for domestic traffic (overall total: 3.3% annual growth rate). <u>Private vehicle</u> travel is expected to increase by 2.2% on a yearly basis over the same period, while <u>intercity bus and rail</u> is anticipated to show negligible growth or continued decline.
Czech Republic		
	Approach to Climate Change	**Target**: Stabilization of net aggregate CO_2, CH_4, and N_2O emissions at 1990 levels by 2000. • The former Czech and Slovak Federal Republic signed the FCCC in Rio in 1992 and was included among signatories undergoing economic transition in Annex 1 to the Convention. Following the separation of the Czech and Slovak Federal Republic, the Czech Republic accepted the terms of the Convention in October 1993.
	GDP 1995 (billion US$)	47.18
	Population 1994 (million)	10.4
	Principal Characteristics and Trends in the Transport Sector	⇒ The transport sector has been strongly affected by <u>recent privatisations</u>; all transport enterprises have been privatised with the exception of the railways. ⇒ <u>Freight transport volume</u> has followed GDP trend, dropping over 20% from the late 1980s to 1994 and stabilizing in 1992. <u>Road</u> freight transport declined overall by 28.4% 1989-1994. <u>Rail traffic</u> dropped 52% in the same period. <u>Inland waterway</u> transport fell 37.5%, levelling off in 1994. <u>Air</u> traffic, however, rose 13% in 1992, averaging 3.7% growth over next few years. ⇒ <u>Public transport share</u> : traditionally around 75 per cent, but under pressure from fast growth in private vehicle ownership. ⇒ <u>Public passenger road traffic</u>: Overall drop of 21% from 1989-1994. <u>Passenger rail transport</u>: 20% decrease since 1992. <u>Passenger air traffic</u>: on the rise: 70% increase in international passenger traffic at Prague Ruzyne airport 1992 to 1994; 23% increase in total passenger traffic 1993-1994.

Denmark		
	Approach to Climate Change	**Target**: For <u>energy sector</u>: 20% reduction in CO_2 emissions by 2005 relative to 1988 (Toronto target); further reductions in emissions of NO_x, SO_2. For the <u>transport sector</u>: stabilization of CO_2 by 2005 at 1988 levels; 25% reduction by 2030; at least 40% reduction in NO_x and HC by 2000, with further reductions thereafter. • These non-binding targets were set out in the Danish action plans for Energy and Transport, which were developed in response to the report of the World Commission on Environment and Development, also known as the "Brundtland Report", and approved by the Danish Parliament in May 1990. • The Danish Parliament ratified the FCCC in December 1993.
	GDP 1994 (billion US$, 1990)	138.9
	Population 1994 (million)	5.2
	Principal Characteristics and Trends in the Transport Sector	⇒ Transport accounted for 7.4% of Danish GDP in 1993 (one-half attributable to road haulage), up from 6.6% in 1984. ⇒ <u>Domestic freight traffic</u>: 94% by road; the rest primarily by short-sea shipping. International <u>freight</u>: 71% by sea in 1992; 23% by road; the rest by rail; <u>maritime transport</u>: 24% increase from 1984 to 1993; <u>rail</u>: 25% drop in freight tonnage between 1986 and 1992. ⇒ <u>Passenger traffic</u>: more than 30% increase in passenger-km travelled from 1984 to 1993, due to rising private car use, which accounts for 75% of total passenger traffic. Railway share is 9% of passenger traffic. ⇒ <u>Overall transport traffic</u> projected to follow rising trend: freight transport to grow by about 40% by 2005; passenger transport by 25% (forecasts based on economic growth estimates).
Finland		
	Approach to Climate Change	**Target**: Stabilization growth in energy-related emissions by the end of the 1990s. • A 1990 Government report on sustainable development identified climate change as the most important environmental issue for the future. • The Finnish Parliament signed the FCCC in April 1994. The Government decided, however, that the stabilization targets would be virtually impossible for Finland to meet; therefore, the above target was adopted.
	GDP 1994 (billion US$, 1990)	124.6
	Population 1994 (million)	5.1
	Principal Characteristics and Trends in the Transport Sector	⇒ Transport <u>expenditure</u> is expected to increase 5% annually from 1995 to 2000 -- a higher growth rate than GDP. As of 2000, transport is projected to grow at approximately the same rate as GDP. ⇒ Total passenger <u>traffic</u> should increase by 30% by 2000. In 1994, <u>public transport</u> accounted for 21% of total passenger-km travelled. ⇒ Total domestic <u>freight traffic</u> is expected to grow in volume by 42% from 1994 to 2000; international and transit traffic by 30%. ⇒ <u>Road traffic</u> grew in kilometres by 55% from 1980 to 1990. <u>Urban</u> road traffic accounts for 94% of total domestic passenger-kilometres and is projected to grow by 2.5% annually from 1995 to 2000. Road haulage predominates <u>freight</u> transport in Finland, carrying 65.1% of domestic traffic in tonne-km (3.7% of international tonnage).

Finland cont'd		⇒ Rail transport handles 64% of long-distance passenger travel. By 2005, total passenger volume is predicted to grow significantly in part due to the introduction of high-speed trains. Railways account for 26.2% of tonne-km in domestic freight traffic; 11.5% of total international freight tonnage. Rail remains the principal mode of transport for raw materials and heavy industry output. Traffic with Russia and the Baltic countries has led to a substantial increase in East-West rail traffic. ⇒ Maritime transport accounts for 85% of international freight traffic but only 8.7% of domestic traffic.
France	Approach to Climate Change	**Target:** Per capita stabilization of net emissions at below 2 tonnes of carbon (equivalent to 7.3 tonnes of CO_2) by 2000, approximately 10% above the 1990 per capita emissions level. • From 1980 to 1990, France succeeded in reducing CO_2 emissions by 23%. It is for this reason that the above target was chosen. • Energy-related CO_2 emissions in France are among the lowest of the OECD countries, largely due to the predominance of nuclear and hydropower in overall energy supply. • The French Government ratified the FCCC in March 1994.
	GDP 1994 (billion US$, 1990)	1 235.4
	Population 1994 (million)	58.3
	Principal Characteristics and Trends in the Transport Sector	⇒ Transport's value added in GDP during the period 1986 to 1995 grew 34% in current francs, despite a 60% drop in maritime transport. ⇒ The current average annual growth rate for freight traffic is 2.1%; for passenger transport, 2.4%. ⇒ Road freight transport excluding transit grew 56% from 1986 to 1995 and is expected to increase 2.4% on an average annual basis from 1992 to 2015. Rail freight traffic dropped 23% from 1986 to 1995, and is projected to increase on an average annual basis by 0.8% from 1992 to 2015 (with an average increase in price of 8% during the period). Combined transport rose 62% from 1986 to 1995, and should continue to grow 5% each year on average from 1992 to 2015. Internal navigation dropped 26% from 1992 to 1995, and should have an average annual growth of 0.8% from 1992 to 2015 (with an average increase in price of 10% during the period). Air freight transport grew 53% from 1985 to 1994 and is expected to grow on average annually 4.9% from 1993 to 2015. ⇒ Passenger mobility has increased by around 24% over the last 10 years in France, with air transport registering the highest growth rate: 48.2% from 1986 to 1995. Passenger road transport is expected to rise by 2.5% annually from 1992 to 2015. Private vehicle traffic increased 29% in passenger-km from 1986 to 1995. In this same period, rail transport declined by around 7% in passenger-km, although the almost one-month-long rail strike at the end of 1995 is probably responsible for a significant part of the decrease. Projections for annual growth in passenger rail from 1995 to 2015 are 1.7%, with prices remaining stable. Overall, passenger use of both urban and inter-city rail transport is in decline.

Germany		
	Approach to Climate Change	**Target**: 25-30% reduction in gross CO_2 emissions by 2005 based on 1987 levels. No official target has been set for other non-Montreal Protocol GHG; the goal is to reduce overall warming effect of all GHG by 50% from 1987 to 2005. • Germany ratified the FCCC in December 1993.
	GDP 1994 (billion US$, 1990)	1 752.6
	Population 1994 (million)	81.4
	Principal Characteristics and Trends in the Transport Sector	⇒ Freight traffic: Road freight transport predominates freight traffic despite efforts to promote other inland forms of freight transport. There has, however, been a slight increase in tonnage transported via rail and inland waterways since 1993. Recent forecasts predict an approximate 45% growth in freight traffic up to 2010. HGV traffic is expected to grow by over 100% from 1988 to 2010, according to the latest estimates. Rail traffic in tonnes-km dropped significantly from 1991 to 1993, increasing slightly from 1993 to 1994. Projections have been made for a 66% growth rate in rail traffic from 1993 to 2010; however, the Government sees this as possible only in the case of a sustained rail-oriented policy. Combined transport increased by 19% from 1993 to 1994, having shown little change for 10 years. The German Government hopes to see combined transport volume double by 2010. Despite a slight decline in tonnage carried from 1990 to 1993, recent forecasts predict 60% growth in inland waterway traffic from 1993 to 2010. Air freight traffic increased by 45% in tonnes-km and 58% in tonnage carried between 1985 and 1993, supporting Government predictions of 151% air traffic growth between 1988 and 2010. Transit traffic registered an increase in flow of 2 million tonnes during the period 1988 to 1993. Government forecasts predict flows of 8.9 million tonnes by 2010. ⇒ Passenger traffic: The modal split has been relatively stable since 1991. Despite incentives to use public transport, the private car was responsible for 81% of all passenger traffic and 82% of passenger-km in 1994. The number of passengers carried by bus and coach has jumped 40% in less than 10 years. Recent rail traffic projections for 1988 to 2010 have been revised downward from 40% to 30%, assuming a strong rail transport development policy, while passenger traffic overall is expected to grow 42% from 1988 to 2010. Estimates for road transport growth are 32% to 43% in the same period. International passenger air traffic -- including scheduled and charter flights -- saw a veritable explosion from 1985 to 1990. Since 1993, however, charter flight traffic has tapered somewhat.
Hungary		
	Approach to Climate Change	**Target**: Stabilization of CO_2 emissions by 2000 at 1985-1987 average emissions level: • Hungary ratified the FCCC in February 1994. Its target is a reflection of the flexibility provided by the Convention as concerns baseline periods for countries with economies in transition. • The achievement of this target and possible further emissions reductions will depend in part on economic growth and improvements in energy efficiency as well as economic and technical co-operation with OECD countries.
	GDP 1994 (billion forints, 1991)	2 478.8
	Population 1994 (million)	10.3

Hungary cont'd	Principal Characteristics and Trends in the Transport Sector	⇒ Freight transport is expected to grow at a moderate rate through 2000, with international and transit traffic increasing at an above-average rate. The railways are expected to lose traffic to road haulage up to 2000, at which time growth in road traffic is projected to taper off due to increasing use of multi-modal transport. Road haulage will continue to predominate after 2000, however.
		⇒ Private passenger transport is expected to increase from 51 billion passenger-km in 1994 to 54-55 billion in 2010. Since the early 1960s, the number of private cars has steadily increased, rising from 30 000 in 1960 to 2.1 million in 1993. Forecasts predict 2.2 million cars in 1995 and 2.5 to 3 million in 2000 depending on economic growth. Public transport will, according to predictions, remain the principal mode of passenger travel until 2000, but its market share will decrease: road and rail public transport should retain their current market share of 57%; internal waterways continue to carry the most leisure travellers; air transport is expected to grow by 60% to 80%; and particularly strong growth is anticipated for international air traffic.
		⇒ Infrastructure building is a current focus of transport sector development, particularly that of the road network. Five major motorways radiating out from Budapest are presently in various stages of development.
		⇒ The Parliament recently adopted a new transport plan, the objectives of which are to:
		  1. promote European integration
		  2. co-operate with neighbouring states to achieve balanced development
		  3. protect human life and the environment
		  4. ensure proper regulation and efficiency of the transport market.

Ireland		
	Approach to Climate Change	**Target**: Limitation of gross CO_2 emissions to 20% growth in 2000 based on 1990 levels. Net CO_2 emissions are expected to grow by 11% between 1990 and 2000. • Ireland developed a CO_2 abatement strategy to contribute to the realisation of the joint EU commitment to stabilize CO_2 emissions at 1990 levels by 2000. The measures included in the strategy will not, however, be sufficient to stabilize Ireland's CO_2 emissions by 2000; therefore, the above target was established. • Ireland ratified the FCCC in April 1994.
	GDP 1994 (billion US$, 1990)	52.6
	Population 1994 (million)	3.6
	Principal Characteristics and Trends of the Transport Sector	⇒ As an island, Ireland is increasingly dependent on its ports and airports for trade. ⇒ Road transport predominates both inland passenger and freight traffic. ⇒ 89% of total freight traffic is carried by road transport. Numbers of commercial vehicles (over 8 tonnes) rose at an average annual rate of 2.8% from 1989 to 1993, despite significant drops in 1992 and 1993. Maritime transport remains the principal mode of international freight traffic, except as concerns light freight or goods with a high added value. Freight traffic at Dublin, Cork and Shannon airports is expected to increase by 6% per year from 1994 to 1999. ⇒ As concerns passenger traffic: public and private road transport accounts for 96% of domestic passenger travel. Rail transport for both passenger and freight traffic dropped 1% in 1994 relative to 1993; however planned investment in rail is expected to lead to a significant increase in passenger traffic: from 7.9 million passengers in 1993 to 8.7 million in 1999.

Italy		
	Approach to Climate Change	**Target**: Italy is committed to the European Community (now EU) target for stabilization of CO_2 emissions at 1990 levels by 2000. Its national target is to stabilize net CO_2 emissions by 2000 at 1990 levels. • Italy signed the FCCC in Rio in 1992 and ratified the Convention in December 1993.
	GDP 1994 (billion US$, 1990)	1 127.2
	Population 1994 (million)	57.2
	Principal Characteristics and Trends of the Transport Sector	⇒ Italy is experiencing a high increase in mobility, with shifts towards road transport for both freight and passenger traffic. Total road traffic rose 132% between 1970 and 1991, whereas GDP increased by 83%. ⇒ Total land-based freight traffic has risen by 150% over the last 20 years, with road transport accounting for more than 80% for long and medium distances. Combined rail-road transport accounted for 6.6% of total rail traffic in 1990. Also that year, 342 million tonnes were transported in and out of Italy, 63% by sea and 21% by road. Over the last two decades, transalpine traffic rose by 119%. Road traffic between Austria and Switzerland accounts for 54% of the road traffic. Growth in road freight traffic with France and Austria has been particularly high over the last 20 years, totalling 641% and 416% respectively. ⇒ As concerns passenger traffic, the number of vehicles per capita has risen by 159% over the last two decades, reaching 49 cars per 100 inhabitants in 1991. This increase in vehicle stocks and car ownership is among the highest in OECD countries. Total transport energy use has more than doubled since 1970, accounting for approximately 30% of total final energy consumption in 1994. Road transport is responsible for 90% of transport energy use. ⇒ In urban areas, local road traffic accounts for around 60% of national road freight traffic and more than 80% of private passenger traffic. Modal shift is strongly in favour of private car travel (OECD, 1994a).
Japan		
	Approach to Climate Change	**Target**: Per capita stabilization of gross CO_2 emissions at 1990 levels by 2000. Efforts to be made to stabilize total emissions beyond 2000 at approx. the same level as in 1990. • Japan signed the FCCC in Rio in 1992 and ratified the Convention in May 1993. • Japan's policy as concerns global warming is based on three elements: "formation of an environmentally sound society, compatibility with a stable development of the economy, and international co-ordination." (IEA/OECD, 1994).
	GDP 1994 (billion US$, 1990)	3 139.5
	Population 1994 (million)	124.9
	Principal Characteristics and Trends of the Transport Sector	⇒ Between 1970 and 1990, Japan's passenger traffic doubled and freight traffic increased by 60% while GDP grew by 133%. ⇒ Compared with other OECD countries, 1990 per capita road traffic was low in Japan; however, trends are showing a considerable modal shift in favour of road transport. The estimated average growth of road traffic in the short term is 2.6% based on economic growth of 4%.

Japan cont'd		⇒ Current <u>modal share</u> for both freight and passenger transport is a relatively favourable 50:50 split between public and private transport. Railways carry 35% of the public transport share in passenger travel. As concerns freight transport, shipping and rail account for 45% and 5% respectively. This modal split is under pressure, however, from rapidly growing shares of car and truck traffic. Indeed, the <u>number of road vehicles</u> more than tripled between 1970 and 1990, while the passenger car fleet rose from 6 million to 32 million. <u>Car ownership</u> remains low in comparison to other OECD countries; however, the size and power of new cars as well as the number of vehicles equipped with automatic transmission and air conditioning have increased recently, and as a result, fuel consumption of the fleet is on the rise (OECD, 1994b).
Latvia		
	Approach to Climate Change	**Target**: Latvia has not adopted a CO_2 target but has stated that it will try to stabilize gross anthropogenic CO_2 emissions. Latvia has also committed to stabilize emissions of other greenhouse gases at a level not exceeding 1990 emissions levels. • Latvia signed the FCCC in Rio in 1992 and ratified the Convention in March 1995.
	GDP 1995 (million lats, 1993)	1 477
	Population 1994 (million)	2.5
	Principal Characteristics and Trends of the Transport Sector	⇒ The share of the transport and communications sector in GDP more than doubled between 1991 and 1994, from 7% to 19.4%. This substantial increase can be attributed in large part to a decline in industry and farming, especially since volumes of both freight and passenger traffic have actually dropped since 1990. ⇒ Regarding <u>freight</u> transport, road freight traffic accounted for less than 20% of total tonne-km in 1990 in the Baltic States. Volume transported in tonnes fell by a factor of 10 between 1990 and 1994. Over the next 20 years, however, road freight traffic is expected to quadruple, as 50% of the current rail traffic volume shifts to road and in response to increased demand for freight traffic between the Nordic countries and Central and Eastern Europe. (This also applies to Lithuania, below). In 1994, Latvian Railways recorded a 4% drop in tonne-km transported by rail relative to the previous year and a 43% drop since 1991. <u>Air</u> freight transport has been stable at approximately 3 000 tonnes, a relatively small share in overall traffic. ⇒ As concerns <u>passenger traffic</u>, <u>car ownership</u> per 1 000 inhabitants is swiftly rising, registering an 87% increase from 1980 (65/1 000) to 1992 (122/1 000). By 2015, car ownership will have risen by 115% to 326 per 1 000, according to medium-term economic growth forecasts. Private cars dropped in number from 1993 to 1994 due to scrapping programmes and modifications to the vehicle registration system. Passenger <u>rail</u> traffic fell by 66% in passenger-km from 1990 to 1994, but seemed to have stabilized somewhat between 1993 and 1994.
Lithuania		
	Approach to Climate Change	**Target**: National greenhouse gas target not yet adopted. • Lithuania signed the FCCC in Rio in 1992 and ratified the Convention in March 1995.
	GDP 1994 (billion litas, 1993)	11.2
	Population 1994 (million)	3.7

Lithuania cont'd	Principal Characteristics and Trends of the Transport Sector	⇒ Lithuania is an important cross-roads for transit traffic from the Baltic countries to Central and Eastern Europe : there are two principal transit corridors : a North-South corridor stretching from the northern tip of Europe to the countries in CEE ; and an East-West corridor linking the Baltic ports of Kaliningrad and Klaipeda to Belarus and Russia. ⇒ Transit road traffic on the North-South Via Baltica grew by more than 500% from 1988 to 1992 : in 1994, average traffic totalled 2 500 to 12 000 vehicles per day depending on the section of the route. On the Lithuanian side of this corridor, traffic density is lower, (1 500 to 5 000 vehicles per day) but is expected to double by 2000. On the East-West corridor, 15.1 million tonnes of goods were transported westwards by rail in 1992, compared with 7.6 million tonnes eastwards. ⇒ Road freight traffic trends closely resemble those in Latvia (see above). ⇒ As concerns maritime transport, freight entering and leaving the port of Klaipeda is primarily transit traffic : in 1994, 77% of exports and 75% of imports passing through the port were in transit, primarily to and from countries in the EU and the Newly Independent States. Maritime passenger traffic has risen considerably : In 1993, 18 500 passengers had used the port of Klaipeda; by 1994, the number of passenger had gone up to 55 000. By 2000, 250 000 to 300 000 passengers are expected. ⇒ Car ownership per 1 000 inhabitants has doubled in the last decade. Vehicle numbers totalled 159 per 1 000 people in 1993, a 20% increase from 1991.
Netherlands		
	Approach to Climate Change	**Target** : Stabilization of gross CO_2 emissions at 1989-90 levels by 1994-95; 3-5% reduction in gross CO_2 emissions at 1989-90 levels by 2000. (3% reduction is targeted -- 5% depending on international developments and opportunities. For road transport: stabilization of CO_2 emissions by 1995 at 1989-1990 levels and 11% reduction by 2000 based on 1990 levels. Other greenhouse gases: 20-25% reduction based on 1989-90 levels by 2000 on a gas-by-gas basis. • The Netherlands signed the FCCC in Rio in 1992 and ratified the Convention in December 1993.
	GDP 1994 (billion US$, 1990)	304.6
	Population 1994 (million)	15.4
	Principal Characteristics and Trends of the Transport Sector	⇒ Transport represented 7.1% of employment in 1990 and 6.4% of GDP in 1992. ⇒ The Netherlands is the leading transport and distribution country in Europe. Dutch carriers account for approx. one-third of international road transport in EU countries and almost half of the freight carried by water, primarily on the Rhine. ⇒ Freight transport (excluding air traffic) increased by approximately 50% from 1970 to 1990, road transport more than doubling to 35 billion tonne-km in 1990, while rail traffic slipped to just slightly more than 3 billion tonne-km. Inland shipping grew during this period from 30.7 to 35.7 billion tonne-km. ⇒ Passenger traffic grew by 97% from 1970 to 1992, while GDP increased by 67% during the same period. Rail transport surged after 1990 with the introduction of a seasonal pass for students. Bicycle use represented 8.6% of total passenger traffic in 1993, and in that same year, private car ownership represented 370 vehicles per 1 000 inhabitants. ⇒ Energy consumption of the transport sector has steadily risen ; road transport accounted for 78% of transport energy consumption, shipping carrying a significant share as well (OECD, 1995c).

New Zealand		
	Approach to Climate Change	**Target**: Primary Objective: Stabilization of net CO_2 emissions at 1990 levels by 2000; Ultimate Objective: Reduction of CO_2 emissions by 20% from 1990 levels by 2000 if based on "no-regrets" measures. • New Zealand signed the FCCC in Rio in 1992 and ratified the Convention in September 1993.
	GDP 1994 (billion US$, 1990)	49.4
	Population 1994 (million)	3.5
	Principal Characteristics and Trends of the Transport Sector	⇒ Energy use in the transport sector increased by 14.8% between 1991 and 1995 (from 133.8 to 153.7 Petajoules). In 1995, 72% of transport sector energy use was for passenger travel and 28% for freight transport. ⇒ As regards passenger transport, energy intensity in overall passenger transport (car, bus, rail and domestic air transport) increased from 2.04 Megajoules (MJ) per passenger-km in 1991 to 2.33 MJ per passenger-km in 1995, an increase of 14.2%. The principal factor influencing the increase in energy intensity has been greater reliance on more energy-intensive means of travel, e.g. domestic air and private vehicle transport. The higher energy intensity of air transport reflects a drop in the average passenger load factor associated with the introduction of domestic competition. As concerns the energy intensity of private vehicle use, per-car occupancy levels have also decreased, while average engine size has increased. ⇒ Total passenger-km travelled increased by 6.1% from 1991 to 1995, with annual passenger-km travelled rising from 13 240 to 13 890 during this period. Of total passenger-km in 1996, private cars accounted for 84.8%, buses and rail 8.9% and domestic air 6.3%. ⇒ As concerns freight transport, energy intensity of total freight transport (road, rail, coastal shipping) remained virtually unchanged from 1991 to 1995 at 2.08 MJ per tonne-km. Reductions in intensity of road freight and coastal shipping transport almost offset the increase in rail freight energy intensity over this period. ⇒ Total freight carried increased by 1.1% from 1991 to 1995. In 1995, 58% of total tonne-km transported was by road and 12% by rail.
Norway		
	Approach to Climate Change	**Target**: Preliminary target: stabilization of gross anthropogenic CO_2 emissions at 1989 levels by 2000 and is continuously analysed in light of technological development, research, and international negotiations and agreements. • Norway signed the FCCC in Rio in 1992 and ratified the Convention in July 1993.
	GDP 1994 (billion US$, 1990)	132.3
	Population 1994 (million)	4.3
	Principal Characteristics and Trends of the Transport Sector	⇒ The transport sector's share in GDP grew by 20% from 1986 to 1991, an increase mainly attributable to the development of maritime activities, the output of which doubled during this period.

Norway cont'd		⇒ <u>Freight</u> traffic remained stable at approximately 300 million tonnes for rail, road and sea transport from 1980 to 1994, with the exception of the period of economic downturn, notably in 1993, when traffic dropped to 280 million tonnes, its 1980 level. Freight level returned to 1991-92 levels in 1994. <u>Road</u> haulage remains the principal mode, carrying 79% of the tonnage transported in 1994. <u>Domestic rail</u> transport is rising for long distances, but tonnage has dropped by half over the last 15 years. <u>International rail</u> has resisted better. Despite being overtaken by road transport, <u>domestic maritime freight flows</u> are still used for long-distance transport (45% share, same as road). ⇒ <u>As concerns passenger traffic</u>, <u>public road transport</u> still holds an important share, despite rising car use. <u>Road</u> transport clearly remains, however, the dominant mode for both public and private transport, accounting for over three-quarters of total traffic. <u>Rail</u> traffic has remained fairly stable overall since 1970. In general, rail services have seen traffic increase since measures were implemented in 1992 to make car use more expensive and favour public transport. <u>Car ownership</u> has sharply risen, however, since 1991.
Poland		
	Approach to Climate Change	**Target** : Stabilization of gross greenhouse gas emissions by 2000 based on 1988 levels. • Poland signed the FCCC in Rio in 1992 and ratified the Convention in July 1994.
	GDP 1994 (000 billion old zlotys, 1993)	1 638.8
	Population 1994 (million)	38.5
	Principal Characteristics and Trends of the Transport Sector	⇒ Poland is expected to record a growth in freight and passenger flows of 170% by 2010. ⇒ Forecasts for <u>passenger</u> traffic modal split to 2005 show an increase in the market share of air transport and the car ownership rate and a reduction in the number of rail and bus passengers. Regarding <u>air</u> transport, the number of passengers is rising more rapidly than the distances covered, suggesting that the domestic market accounts for a considerable share of total traffic. Passenger transport by <u>sea</u> is growing: ferry traffic in Polish ports is expected to rise from 690 000 passengers in 1993 to 1 900 000 in 2010. ⇒ Regarding <u>freight</u> traffic, a significant increase in road transport is expected, with rail transport and pipeline traffic in decline. In 1990, <u>road</u> accounted for 29% of tonne-km transported and <u>rail</u> 60.1%; by 1995, road held 38% and rail 51%. Projections are that by 2005, road will predominate with 49.5% of tonne-km and rail 44.1%.
Portugal		
	Approach to Climate Change	**Target** : No national target; recognition of role in EU-wide target (stabilization of greenhouse gas emissions at 1990 levels by 2000). • Portugal signed the FCCC in Rio in June 1992 and ratified the Convention in December 1993.
	GDP 1994 (billion US$, 1990)	69.2
	Population 1994 (million)	9.9

Portugal cont'd	Principal Characteristics and Trends of the Transport Sector	⇒ Portugal's transport sector experienced significant growth in the 1980s. The <u>road</u> vehicle fleet grew by 104% and car ownership rose by 77%. (Portugal still remains well below OECD averages). Total road traffic in vehicle-km rose by 67%, while GDP grew by 31%. Total car fuel consumption grew by 46%, despite a decline in average per-kilometre consumption of 8%. ⇒ <u>Passenger</u> traffic is undergoing a decisive shift from rail to road: total passenger traffic increased by 44% in the 1980s, and road traffic climbed by 55%. From 1985 to 1994, passenger road traffic increased by 73% in passenger-km. ⇒ <u>Freight</u> transport by rail continues to grow, as does road haulage if to a lesser extent (50% in the 1980s). Since 1993, there has been no significant change in rail traffic. Over the next 20 years, rail traffic volumes on the two main railway corridors (Lisbon-Porto, Beira Alta) are projected to double (OECD, 1993).
Romania		
	Approach to Climate Change	**Target** : National target not yet established. • Romania signed the FCCC in Rio in 1992 and ratified the Convention in June 1994.
	GDP 1994 (billion lei, 1993)	20 817
	Population 1994 (million)	22.7
	Principal Characteristics and Trends of the Transport Sector	⇒ A 1995 survey showed that <u>international road</u> traffic accounted for 7% of total road traffic and transit transport 4%. ⇒ <u>Rail freight</u> traffic is benefiting more from the economic recovery than passenger transport by rail, according to a 1995 survey: forecasts show 18% growth in total freight traffic from 1994 to 2000 (99 179 tonnes to 117 000 tonnes, and only 1.8% growth in passenger traffic over the same period (206 920 to 211 000 passengers). ⇒ <u>Combined transport</u> has grown in some areas, especially since 1992; however, volumes of traffic remain below those elsewhere in Europe. <u>Container</u> traffic volumes dropped by 40% in 1993 relative to the previous year, but rose by 20% in 1994. Projections are that this trend will continue, rising by 60% by 2000.
Russian Federation		
	Approach to Climate Change	**Target** : The Russian Federation intends to take actions to reduce CO_2 and other greenhouse gas emissions according to its commitments to the FCCC. No specific stabilization or reduction targets have been set to date. • Russia signed the FCCC in Rio in 1992 and ratified the Convention in December 1994.
	GDP 1994 (billion US$, 1990)	N.A.
	Population 1994 (million)	148.0
	Principal Characteristics and Trends of the Transport Sector	⇒ <u>Medium-term priorities</u> outlined in the State transport policy focus on developing passenger and freight transport with consideration for "social" aspects, as well as improving and developing transport communications, modernising and renewing transport equipment, and increasing the operational safety of transport systems.

Russian Federation cont'd	⇒ The focal point of <u>institutional development</u> in the transport sector concerns <u>privatisation</u>. By the end of 1995, some 3 500 State enterprises, representing 64% of all enterprises in the Ministry of Transport system scheduled for privatisation, had become joint stock and private companies. This was the case for 93% of inland water transport companies, more than 76% of road transport companies, and 65% of maritime transport companies. Privatised enterprises now employ 64% of all workers (1 375 000 people) and perform about 80% of all transport services. ⇒ <u>Tariff policy</u> is aimed at further liberalisation of the transport sector. A free tariff structure has been introduced for road freight transport, inland waterway freight and air freight transport. Tariffs are set by transport enterprises in agreement with freight consignment companies. The same tariff structure is being applied to air passenger transport on regular airlines and to inter-city passenger transport by coach and inland waterway transport.

Slovak Republic		
	Approach to Climate Change	**Target**: 20% reduction in net emissions by 2005 compared to 1988. • The former Czech and Slovak Federal Republic signed the FCCC in Rio in 1992. The Slovak Republic signed the Convention after it became an independent state in May 1993. It ratified the Convention in August 1994.
	GDP 1994 (billion slk koruny, 1993)	388.0
	Population 1994 (million)	5.3
	Principal Characteristics and Trends of the Transport Sector	⇒ <u>Freight transport by rail</u> has been decreasing in volume since 1989. Overall import flows dropped by 23.4%, while export flows declined as well but to a lesser extent - 6.8%. Most seriously affected by the recession were transit traffic flows -- down 61%. A large part of the reason for a particularly sharp decline in 1993 rail traffic was a deterioration in trade relations between the Slovak and Czech Republics, and between the Czech Republic and Hungary and other countries of south-eastern Europe -- transit traffic which generated trade relations for the Slovak Republic. ⇒ <u>Passenger traffic</u> has steadily dropped each year since the mid-1980s, resulting in an overall decrease of around 35%. Passenger numbers slipped from 117.1 million in 1990 to 89.5 million in 1995. Traffic volumes appear to be picking up slightly since 1995, however, as the economy recovers. <u>Rail</u> traffic has been in decline since 1990, notably those travelling at reduced "worker" fares; this is attributable to rising unemployment. ⇒ <u>International and cross border</u> rail traffic with the Czech Republic has been on the rise since 1994. Forecasts for 2000 predict moderate economic growth from 1996 to 1998, accompanied by higher car ownership levels and more rail transport links with border regions. ⇒ As concerns <u>road</u> transport, forecasts indicate a moderate increase to the year 2000. High fuel prices have kept average annual vehicle-km relatively low. The current rail-road modal split, 70% and 30% respectively, may be reversed to a 40%-60% split if current market trends prevail.

Slovenia		
	Approach to Climate Change	• Slovenia signed the FCCC in Rio in 1992 and ratified the Convention in December 1995.
	GDP 1994 (billion tolars, 1992)	1 087.6
	Population 1994 (million)	1.9
	Principal Characteristics and Trends of the Transport Sector	⇒ Both <u>freight and passenger</u> traffic have been in decline since 1985-1986, in spite of a slight recovery in 1994, particularly as concerns freight. ⇒ <u>Freight</u> traffic declined in both tonnage and tonne-km from 1985 to 1994: in 1985, 18 649 tonnes of freight were transported, falling to 5 442 tonnes in 1994, a 71% drop. Likewise, 3 772 tonne-km of goods were hauled in 1985, slipping to 1 935 tonne-km in 1994, a 49% decline. ⇒ <u>Transit</u> traffic accounts for half of freight traffic, with imports accounting for 30% of the total -- three times the percentage of domestic freight. The share of imports is expected to increase from 6.5% to 7.5% in 1998; tonnage will most likely follow suit. ⇒ <u>Rail</u> transport predominates freight flow traffic, with road transport accounting for only 25%. In 1994, rail accounted for 60% of tonnage carried. The number of lorries increased, however, by 20% between 1980 and 1994. ⇒ As concerns <u>passenger</u> transport, rail accounted for 50.2% of total passenger-km in 1994 compared with 46.3% in 1990. The total volume of rail traffic has declined by almost 40% since 1980, however, the biggest drop in number of passengers occurring on the international network. As regards urban and intercity trips, road transport accounted for 53% of passengers carried in 1994, with rail carrying 42.4%. The <u>number of private cars</u> rose by almost 60% between 1980 and 1994, leading to a car ownership ratio of 333 vehicles per 1 000 inhabitants.
Spain		
	Approach to Climate Change	**Target** : Increase in energy-related CO_2 emissions limited to 15% between 1990 and 2000 (revised downward from 25% in December 1995). The target has been set within the context of the EU's global policy towards stabilization of CO_2 emissions at 1990 levels by 2000, considering the particular circumstances of individual countries and the principle of burden-sharing. • Spain signed the FCCC in Rio in 1992 and ratified the Convention in December 1993.
	GDP 1994 (billion US$, 1990)	511.1
	Population 1994 (million)	39.2
	Principal Characteristics and Trends of the Transport Sector	⇒ Forecasts carried out in the early 1990s for both <u>passenger and freight</u> transport show steady growth in road traffic and more uneven growth in rail traffic. ⇒ As concerns <u>domestic freight transport</u>: <u>road</u> traffic accounts for 80% of domestic freight traffic; <u>maritime</u> traffic remains the second largest mode of freight transport in Spain (14%), despite a significant drop in volume between 1993 and 1994 (over 15% decline between 1989 and 1994); <u>rail</u> accounts for 4% and <u>pipeline</u>, 2%. ⇒ <u>International freight transport</u> is dominated by <u>maritime</u> traffic (76%) and <u>road</u> traffic (23%). The share of <u>rail</u> (1%) and <u>air</u> are very small. <u>Air</u> traffic accounted for less than one million tonnes of goods transported in 1994, although air freight volume increased by 40% from 1991 to 1994.

Spain cont'd		⇒ As regards <u>domestic passenger traffic,</u> <u>road</u> traffic is expected to increase by 56% to 308 billion passenger-km between 1989 and 2000. <u>Rail</u> traffic is likewise projected to grow by 50% from 16 billion to 24 billion passenger-km during the same period. Although growth in air transport has been higher than that of any other mode in the early 1990s (+58%), road transport maintains its hold on the highest share of the market (90% in 1994). <u>Rail</u> accounted for 6% of traffic volume in 1994, an only 0.6% increase since 1989. ⇒ <u>International passenger traffic</u> is likewise dominated by road transport. <u>Road</u> traffic volume grew by 65% between 1990 and 1991, accounting for up to 65% of cross-border traffic. <u>Air</u> traffic follows road in modal share (28% of the international passenger transport market) while international <u>rail</u> traffic carries less than 4% of total traffic. Modernisation of the rail network and introduction of high-speed services is expected to revive long-distance rail transport; by 2000, traffic in passenger-km could rise by a factor of 2.5. However, the decline in demand for long-distance rail transport (33% drop from 1989 to 1994) may suggest that this projection is somewhat optimistic. <u>Maritime</u> traffic has remained stable since 1989, accounting for less than 3% of all traffic. ⇒ <u>Domestic freight traffic</u> trends generally follow those of passenger traffic, with road traffic growing steadily and rail progressing at a more uneven rate. ⇒ As concerns <u>international</u> freight transport, overall volume of traffic is split between maritime transport (76%) and road transport (23%). Rail and air shares are either non-existent or insignificant.
Sweden		
	Approach to Climate Change	**Target** : Stabilization of gross greenhouse gas emissions at 1990 levels by 2000; reduction after 2000 based on 1990 levels. 30% reduction in methane emissions from refuse disposal by 2000 based on 1990 levels. • Sweden signed the FCCC in Rio in 1992 and ratified it in June 1993.
	GDP 1994 (billion US$, 1990)	224.6
	Population 1994 (million)	8.8
	Principal Characteristics and Trends of the Transport Sector	⇒ <u>Road</u> traffic: The number of vehicles now averages 400 cars per 1 000 inhabitants. A considerable share of the Swedish vehicle stock is comprised of large, heavy, high-powered cars. Car travel is responsible for about 75% of all personal transport. As concerns road freight, 40% of overall freight volume is transported by lorry; for short-distance haulage, the lorry predominates other modes of transport. ⇒ <u>Railways</u> account for 15% of long-distance passenger transport. Regarding <u>freight</u> transport, railways carry 25% of domestic, long-distance freight, its share having diminished in recent years. ⇒ A significant share of freight is transported by <u>shipping</u>, which accounts for 12% of long-distance domestic goods transport and 50% of long-distance international shipments. <u>Ferry</u> traffic across the Baltic Sea has risen considerably during the 1980s. (Sweden, 1994)

Switzerland		
	Approach to Climate Change	**Target**: Stabilization of gross CO_2 emissions at 1990 levels by 2000; reduction in gross CO_2 emissions after 2000. • Switzerland signed the FCCC in Rio in June 1992 and ratified it in December 1993.
	GDP 1994 (billion US$, 1990)	226.2
	Population 1994 (million)	6.9
	Principal Characteristics and Trends of the Transport Sector	⇒ Freight traffic is projected to continue to grow at an increasingly rapid rate. From 1970 to 1992, tonne-km of freight traffic rose 2.1% per year; over the next two decades, annual growth rates are projected to vary between 2.7% and 3.5% depending on the scenario. Rail freight transport dropped from 1990 to 1993, in large part due to a falling off in North-South transit traffic. As of 1994, the trend turned back upward. Regarding transalpine traffic, tonnage carried by rail and combined transport rose by 2.3% and 2.2% respectively from 1992 to 1994. ⇒ From 1994 to 1995, total rail and road traffic through the Swiss Alps increased by 3%. Rail accounted for 73% of total freight tonnage transiting the Swiss Alps in 1995, 17.9 million tonnes. Combined transport grew around 5% the same year, a smaller increase than in 1994. Road freight traffic fell overall from 1990 to 1993, however transit traffic by road grew substantially in terms of tonnage carried -- by 19.6 % from 1992 to 1994. In 1995, transalpine road traffic rose by 6% and transported 27% of total freight tonnage crossing the Swiss Alps. ⇒ Passenger traffic is expected to increase to 2015, albeit at a less rapid rate than over the past 20 years. In terms of passenger-km, traffic is projected to grow by approximately 40%: from 98 billion passenger-km in 1990 to 135 to 140 billion in 2015. Growth in the share of public transport (rail and road) in passenger traffic has slowed in recent years, however, public transport has been able to maintain and slightly increase its overall market share. Traffic trends to 2015 show a favourable evolution in public transport use. International rail passenger transport, on the rise until 1992, began to fall from 1992 to 1994; however, the share of rail in overall passenger transport is expected to recover, increasing by between 13% and 18% by 2015. Road passenger transport dropped in 1993, most likely due to the economic recession and higher fuel duties. From 1990 to 1993, however, overall private traffic rose by 3.4%. Growth trends in private road transport are projected to continue at a lower rate than other public transport modes.
Turkey		
	Approach to Climate Change	**Target**: Although Turkey recognises the need to reduce greenhouse gas emissions, it feels that countries should share the burden of abatement in ways that reflect their relative level of development. The Government has stated that Turkey's contribution to global CO_2 is negligible; therefore it has not adopted a national greenhouse gas or CO_2 emissions target. • Turkey attended the Rio Convention in June 1992, but has stated that it will not sign the Convention until it is no longer included among developed countries in the annexes to the Convention. Being considered a developed country carries with it responsibilities that Turkey believes are in contradiction with its level of development.
	GDP 1994 (billion US$, 1990)	164.6

Turkey cont'd	Population 1994 (million)	60.6
	Principal Characteristics and Trends of the Transport Sector	⇒ Transport policy and sector projections are laid out in the current (seventh) five-year transport plan, which covers the period from 1995 to 1999. <u>Road</u> transport accounts for 95% of domestic transport and is expected to retain this share, despite a decline in 1994. ⇒ <u>Freight</u> traffic: <u>Rail's</u> share in the freight transport modal split should not exceed 10% during the period 1995-1999, despite earlier projections of a 11.7% share. Rail freight traffic in tonne-km is predicted to increase by 125% from 1989 to 2000, 84% between 1994 and 2000. <u>Domestic rail</u> freight traffic is expected to grow by 10.9% per year, a revision upward from the previous transport plan, which projected 8.9% annually. <u>Road</u> freight dropped by 3% in tonne-km from 1993 to 1994 (97 843 to 95 020 tonne-km). ⇒ <u>Passenger</u> rail traffic within Turkey is forecasted to increase from 3.6 billion passenger-km in 1989 to 5.1 billion passenger-km in 2000. <u>Road</u> transport slipped by 4% in passenger-km during the period 1993 to 1994, but remains by far the predominant mode.
United Kingdom		
	Approach to Climate Change	**Target** : Stabilization of gross greenhouse gas emissions by 2000 at 1990 levels. Specific targets set for different greenhouse gases: as concerns CO_2, the target is to return emissions to 1990 level by 2000. • The UK signed the FCCC in Rio in 1992 and ratified the Convention in December 1993.
	GDP 1994 (billion US$, 1990)	1 009.7
	Population 1994 (million)	58.4
	Principal Characteristics and Trends of the Transport Sector	⇒ <u>Freight</u> transport: Road traffic largely predominates freight traffic with 65% of the market share in 1994, representing 144 billion tonne-km. Behind road is waterway transport, with 52 billion tonne-km and 24% of the market. Rail follows as a distant third, with 13 billion tonne-km and 6% of the market share. Projections give no indication that the supremacy of road transport is at risk. A large share of the anticipated growth in road traffic is expected to come from large trucks (four axles and over); forecasts show an increase in traffic in vehicle-km of 15% to 20% by 2000, and of 97% to 195% by 2025 compared with 1994. Light goods vehicle (LGV) traffic in vehicle-km is projected to increase by 12% to 20% by 2000 relative to 1994 levels, and by 78% to 190% by 2025. The trend in the number of LGVs has closely matched GDP growth and is likely to continue to do so in the future. As concerns rail traffic, the Channel Tunnel is expected to improve growth projections for rail freight transport; however it is unlikely to be sufficient to challenge the predominance of road transport for domestic flows. Rail traffic declined from 1990 to 1995, particularly in the traditionally strong rail markets such as coal traffic, which recorded a 34% drop during this period.

United Kingdom cont'd		⇒ <u>Passenger</u> traffic: <u>Private car</u> traffic is expected to rise by 11% to 19% until 2000, and then by 57% to 87% thereafter to 2025 relative to 1994 levels. <u>Car ownership</u> represented 378 vehicles per 1 000 inhabitants in 1992. A <u>low-growth</u> scenario forecasts that this rate will grow to 411 in 2000 and 529 by 2025. <u>High-growth</u> projections estimate 428 cars per 1 000 inhabitants in 2000 and 579 in 2025. As concerns <u>rail</u> transport: both the number of railway trips made and passenger-km trends were in decline from 1990/1991 to 1994/1995, recovering slightly in 1995/1996. The number of trips dropped by 8% from 1990 to 1995, in this period, from 762.4 million to 702.2 million, recovering by 2% to 718.7 million trips by 1996. In the same way, passenger-km slipped from 33.2 billion in 1990 to 28.7 billion in 1995, regaining 5% by 1996 to log 30 billion passenger-km.
United States		
	Approach to Climate Change	**Target**: Stabilization of aggregate global warming potential of net emissions of all greenhouse gases at 1990 levels by 2000. • The United States signed the FCCC in Rio in 1992 and ratified it first among OECD member countries in October 1992.
	GDP 1994 (billion US$, 1990)	6 027.1
	Population 1994 (million)	260.6
	Principal Characteristics and Trends of the Transport Sector	⇒ In 1991, motor vehicles were driven 3 500 billion kilometres, more than in all other OECD countries combined and double the distance travelled two decades ago. Per capita vehicle-km travelled was almost twice the OECD average. And car ownership ranked the highest among OECD countries at 570 per 1 000 inhabitants. ⇒ <u>Passenger transport</u> accounted for 6 700 billion passenger-km in 1992, an 86% increase from 1970. GDP growth in the same period was 74%. Per capita travel averaged 25 000 km per year, approximately 90% of which was by motor vehicle. Surface <u>public transport</u> carried only 3% of total passenger traffic. <u>Air</u> transport, accounting for 9% of passenger traffic, is the fastest growing mode. ⇒ <u>Freight</u> traffic totalled approximately 4.8 billion tonne-km in 1992 excluding coastal waterway transport. Railways carried 37%, trucks -- 29%, inland waterways -- 15%, oil pipelines -- 19% and air -- less than 1%. Inter-city freight transport grew by approximately 50% between 1970 and 1990. Rail's share remained relatively stable, and the share of trucks grew gradually during this period (OECD, 1996c).
European Union		
	Approach to Climate Change	**Target**: Stabilization of gross CO_2 emissions at 1990 levels by 2000. • The European Community[5] signed the UN FCCC at the Rio Conference in 1992 and ratified it on 21 December 1993. It is the only "regional economic integration organisation" that is Party to the FCCC and is included in the two Annexes to the Convention. The EU is committed to adopt policies and measures to stabilize greenhouse gas emissions at 1990 levels, jointly or individually, by 2000. As concerns CO_2 emissions, the EU decided in an October 1990 joint Council meeting that CO_2 emissions in the Union should be stabilized at their 1990 levels by 2000. This target is jointly shared by the Union's Member countries and will not have to be achieved by each country individually; in other words, growth in CO_2 emissions in EU countries that have below average CO_2 emissions must be offset by emissions reductions in other EU countries.

European Union cont'd		• In light of the decision of the Second Conference of the Parties in Berlin, March-April 1995, the EU has additionally committed to limit and reduce emissions beyond 2000 (EC, 1995).
	GDP 1994 (billion US$, 1990)	7 020.2
	Population 1994 (million)	371
	Principal Characteristics and Trends of the Transport Sector	⇒ Transport has significantly contributed to economic growth in the EU, enabling considerable economies of scale in production and leading to increased competition. The transport sector is responsible for approximately 7-8% of GDP and 4-5% of the salaried workforce (White Paper on the Common Transport Policy, COM (92)494 final 2.12.1992 cited in EC, 1996). ⇒ The sector has undergone and is expected to experience significant growth. Projected increases in vehicle-km from 1990-2000 are : 17% for passenger cars; 26% for LGVs; 13% for two-wheelers. ⇒ According to the International Civil Aviation Organisation, air traffic is expected to grow in Europe (not including the Newly Independent States) by 5.2 per cent annually until 2003 (EC, 1996).

3.2 Summary of emissions data

The following table is designed to summarise principal emissions information provided from each of the countries which responded to the questionnaire. For each country, base year data and projected growth in emissions to 2000 and 2010 is given for road transport and the transport sector. Transport's share in overall emissions is also indicated. A Notes section is included where additional information was available and appropriate. It should be kept in mind that percentage changes in emissions are derived from the questionnaire responses and are approximate. Complete comparative tables of the CO_2 emissions data provided by the countries are found in Annex 4.

Table 3: **Summary of CO_2 emissions in ECMT countries**

	BASE YEAR 1990 unless otherwise indicated (1 000 tonnes)	REFERENCE CASE Projections 2000, 2010 (no additional measures since base year)	STATUS QUO Projections 2000, 2010 (with only measures implemented since base year)	FUTURE Projections 2000, 2010 (with measures implemented or planned)
Austria	• Road emissions : 13 280 • Sector emissions : 16 161 • Transport share in total : 27.1%	• Road emissions to 2000 : +19% to 2010 : +20% • Sector emissions to 2000 : +20% to 2010 : +21% • Transport share in total: 2000 : 29.7% 2010 : 27.4% Modelling exercise - 1995	Not provided	• Road emissions to 2000 : +19% to 2010 : + 9% • Sector emissions to 2000 : +20% to 2010 : +12% • Transport share in total : 2000 : 29.7% 2010 : 25.4%
	Average change in annual transport emissions 1990-95: + 3.6%.			
	Notes: Data projections take into account that in 2003, road pricing will be introduced on motorways and in 2010 on all Austrian highways. Vignette to be introduced 1-97.			
Belgium	• Road emissions : 21 000 • Sector emissions : 22 000 • Transport share in total : 20%	• Road emissions to 2000 : +38% to 2005 : +57% • Sector emissions to 2000 : +36% to 2005 : +54% • Transport share in total : 2000 : 24% 2005 : 27%	Not provided	• Road emissions to 2000 : +26% to 2005 : +43% • Sector emissions to 2000 : +25% to 2005 : +36% • Transport share in total : 2000 : 24.5 % 2005 : 25.2%
	Average change in annual transport emissions 1990-95: + 4.4%.			
	Notes: For all cases: Rail, Shipping and Aviation data combined; no data for motorcycles and LCVs.			
Canada	• Road emissions : 117 800 • Sector emissions : 140.100 • Transport share in total : 31.0%	Not provided (see Notes below)	• Road emissions to 2000 : +11% to 2010 : +24% • Sector emissions to 2000 : +11% to 2010 : +23% • Transport share in total : 2000 : 32% 2010 : 32%	Not provided
	Average change in annual transport emissions 1990-95: +5%.			
	Notes: Canada noted that the Canadian Reference case corresponds best to the Status Quo case of the questionnaire, in that all current energy and related policies are held constant over the projection period, and no new initiatives are included.			
Czech Republic	• Road emissions : 6 840 • Sector emissions : 7 926 • Transport share in total : 4.7%	• Road emissions to 2000 : + 65% to 2010 : +106% • Sector emissions to 2000 : + 68% to 2010 : +108% • Transport share in total : 2000 : 11.6% 2010 : 14.5%	• Road emissions to 2000 : +42% to 2010 : +70% • Sector emissions to 2000 : +41% to 2010 : +70% • Transport share in total : 2000 : 9.7% 2010 : 11.8%	• Road emissions to 2000 : +31% to 2010 : +36% • Sector emissions to 2000 : +29% to 2010 : +33% • Transport share in total : 2000 : 8.9% 2010 : 9.2%
	Average change in annual transport emissions 1990-95: +5%.			
	Notes: Energy system model (EFOM) compared with fuel data; internal measures effectiveness analysis (RMEA) used. Influence of GNP growth and demographic evolution included. Improvement in methodology signalled since 1993 questionnaire. Estimates for 2010 based on assumption that increase rate 2005-2010 approx. 50% less than rate 2000-2005.			

	BASE YEAR 1990 unless otherwise indicated (1000 tonnes)	**REFERENCE CASE** Projections 2000, 2010 (no additional measures since base year)	**STATUS QUO** Projections 2000, 2010 (with only measures implemented since base year)	**FUTURE** Projections 2000, 2010 (with measures implemented or planned)

Denmark	• Road emissions : 9 239
	• Sector emissions : 10 236
	• Transport share in total : 17%
	
	Base year = 1988
		• Road emissions
		  to 2000 : +14%
		  to 2005 : +21%
		  to 2010 : +25%
		• Sector emissions
		  to 2000 : +10%
		  to 2005 : +16%
		  to 2010 : +19%
		• Transport share in total :
		  2000 : 21%
		  2005 : 22%
		  2010 : 22%
			Not provided	• Road emissions
				  to 2005 : +3%
				• Sector emissions to
				  2005 : -0.1%
				  Transport share in total :
				  2005 : 21%
				
				Future projections account for the Danish Government's *Actio100n Plan for Reduction of CO$_2$ Emissions from the Transport Sector.*

	Average change in annual transport emissions 1990-95: +1.3%. **Notes:** Danish target is to stabilise emissions at 1988 level by 2005 ; therefore Future Case is for 1988 and 2005. Future projections account for the Danish Government's *Action Plan for Reduction of CO$_2$ Emissions from the Transport Sector.*			

Finland	• Road emissions : 11 200
	• Sector emissions: 15 600
	• Transport share in total : 24%
		• Road emissions
		  to 2000 : +20%
		  to 2010 : --
		• Sector emissions
		  to 2000 : +20%
		  to 2010 : --
		• Transport share in total :
		  2000 : 25%
		  2010 : --
			• Road emissions
			  to 2000 : +0.9%
			  to 2010 : +3.5%
			• Sector emissions
			  to 2000 : +6%
			  to 2010 : +9%
			• Transport share in total :
			  2000 : 22%
			  2010 : 24%
				Not provided

France	• Road emissions : 111 500 • Sector emissions: 132 800 • Transport share in total : 34.7%	Not provided	Not provided	Not provided
	Average change in annual transport emissions 1990-95: +1.5%. **Notes :** Data derived from the annual report of the *Commission des Comptes des Transports de la Nation* on transport energy use and not from the greenhouse gas inventory for the FCCC. Data thus differs between the two reports ; notably as concerns differences in methodology ; the data presented here does not include overseas *départements* ; emissions are corrected for climatic variations.			
Germany	• Road emissions : 150 000 • Sector emissions : 159 000 • Transport share in total : 15.7%	Not provided	Not provided	Not provided
	Average change in annual transport emissions 1990-95: +2.4% **Notes :** Data taken from a draft of Germany's second communication to the FCCC.			

Hungary	• Road emissions : 8 131.6
	• Sector emissions : 10 360.4
	• Transport share in total : 13.8%
		• Road emissions
		  to 2000 : +31%
		  to 2010 : --
		
		No projections provided for 2000 because of economic uncertainty.
			Not provided	Not provided

	Notes : It is noted that CO$_2$ emissions in 2000 and 2010 will depend on the rate of economic development and changes in energy demand.			

	BASE YEAR 1990 unless otherwise indicated (1 000 tonnes)	**REFERENCE CASE** Projections 2000, 2010 (no additional measures since base year)	**STATUS QUO** Projections 2000, 2010 (with only measures implemented since base year)	**FUTURE** Projections 2000, 2010 (with measures implemented or planned)

Ireland	• Road emissions : 4 715
	• Sector emissions : 6 057
	• Transport share in total : 19%
		• Road emissions
		  to 2000 : +25%
		  to 2010 : --
		• Sector emissions
		  to 2000 : +25%
		  to 2010 : --
		• Transport share in total :
		  2000 : 19.7%;
		  2010 : --
			Not provided	Not provided

	Average change in annual transport emissions 1990-95: +5.2%			
	Notes : Road transport not disaggregated.			

Italy	• Road emissions : 912
	• Sector emissions : 1 098
	• Transport share in total : 27.4%
		• Road emissions
		  to 2000 : +21%
		  to 2010 : --
		• Sector emissions
		  to 2000 : +10%
		  to 2010 : --
		• Transport share in total :
		  2000 : 26%
		  2010 : --
			Not provided	Not provided

Japan	• Road emissions : -- • Sector emissions : 215 000 • Transport share in total : 18.3%	Not provided	Not provided	Not provided
	Average change in annual transport emissions 1990-94: +2.9%			

Latvia	• Road emissions : 3 826.1
	• Sector emissions : 5 662.5
	• Transport share in total : 24%
		• Road emissions
		  to 2000 : - 18%
		  to 2010 : --
		• Sector emissions
		  to 2000 : - 41%
		  to 2010 : --
		• Transport share in total :
		  2000 : 21.4%
		  2010 : --
			Not provided	Not provided

	Average change in annual transport emissions 1990-95: -16%			

Lithuania	• Road emissions : 3 681
	• Sector emissions : 4 498
	• Transport share in total : 12.3%
		• Road emissions
		  to 2000 : - 7%
		  to 2005 : - 9%
		• Sector emissions
		  to 2000 : - 8%
		  to 2005 : +13%*
		• Transport share in total :
		  2000 : --
		  2010 : --
			• Road emissions
			  to 2000 : - 4%
			  to 2005 : +38%
			• Sector emissions
			  to 2000 : - 3%
			  to 2005 : +13%*
			• Transport share in total :
			  2000 : --
			  2010 : --
				• Road emissions
				  to 2000 : - 4%
				  to 2005 : +82%
				• Sector emissions
				  to 2000 : - 3%
				  to 2005 : +13%*
				• Transport share in total :
				  2000 : --
				  2010 : --

	* Data taken from National Communication to FCCC by Ministry; other data from other sources.			

	BASE YEAR 1990 unless otherwise indicated (1000 tonnes)	**REFERENCE CASE** Projections 2000, 2010 (no additional measures since base year)	**STATUS QUO** Projections 2000, 2010 (with only measures implemented since base year)	**FUTURE** Projections 2000, 2010 (with measures implemented or planned)

Netherlands	• Road emissions:
	   23 800
	• Sector emissions:
	   27 700
	• Transport share in total:
	   15%
		• Road emissions
		   to 2000: --
		   to 2010: +51%
		• Sector emissions
		   to 2000: --
		   to 2010: +51%
		• Transport share in total:
		   2000: --
		   2010: --
			• Road emissions
			   to 2000: --
			   to 2010: +29%
			• Sector emissions
			   to 2000: --
			   to 2010: +30%
			• Transport share in total:
			   2000: --
			   2010: --
				• Road emissions
				   to 2000: --
				   to 2010: +8%
				• Sector emissions
				   to 2000: --
				   to 2010: +10%
				• Transport share in total:
				   2000: --
				   2010: --

	Average change in annual transport emissions 1990-95: +2.8%			

New Zealand	• Road emissions:
	   --
	• Sector emissions:
	   11 161
	• Transport share in total:
	   40%
		Not provided	• Road emissions:
			   From 2000 to 2010:
			      +22%
			• Sector emissions
			   to 2000: +40%
			   to 2010: +69%
			• Transport share in total:
			   2000: 43%
			   2010: 44%
			
			For road emissions, base year not included in modelling exercise. For sector emissions, base year is 11 161. (See notes)
				Not provided

	Average change in annual transport emissions 1990-95: +4%			
	Notes: Status Quo emissions case from Ministry of Commerce modelling programme conducted in late 1995. The Ministry specified that it is a "central" scenario as opposed to an expectation of future emissions. It is based on 3% annual growth in GDP, with oil prices rising to US$25 bbl by 2005 stabilising thereafter, and current margins on crude prices maintained for petroleum products. Databases used for base year and projections are different; therefore caution is advised in interpreting growth.			

Norway	• Road emissions:
	   8 000
	• Sector emissions:
	   11 400
	• Transport share in total:
	   35.2%
		Not provided	• Road emissions
			   to 2000: -0.06%
			   to 2010: +20%
			• Sector emissions
			   to 2000: --
			   to 2010: --
			• Transport share in total:
			   2000: --
			   2010: --
				Not provided

	Average change in annual transport emissions 1991-94: +2.1%			
	Notes: The Norwegian strategy on CO_2 emissions could be characterised as "business as usual" since 1991, according to the Ministry of Transport; the Norwegian Green Tax Commission assumed unchanged tax level on fuels in real prices.			

Poland	• Road emissions:
	   20 016
	• Sector emissions:
	   28 498
	• Transport share in total:
	   --
	
	Base year = 1995
		• Road emissions
		   to 2000: +25%
		   to 2010: +76%
		• Sector emissions
		   to 2000: +20%
		   to 2010: +61%
		• Transport share in total:
		   2000: --
		   2010: --
			Not provided	• Road emissions
				   to 2000: +15%
				   to 2010: +50%
				• Sector emissions
				   to 2000: +13%
				   to 2010: +42%
				• Transport share in total:
				   2000: --
				   2010: --

	Notes: Data from the questionnaire is based on the results of the *Polish Country Study to Address Climate Change*, carried out under a co-operative agreement between the governments of Poland and the United States (January 1996). The results of the study are currently under verification; therefore numbers are to be considered as preliminary.			

	BASE YEAR 1990 unless otherwise indicated (1 000 tonnes)	REFERENCE CASE Projections 2000, 2010 (no additional measures since base year)	STATUS QUO Projections 2000, 2010 (with only measures implemented since base year)	FUTURE Projections 2000, 2010 (with measures implemented or planned)
Portugal	• Road emissions : 9 413 • Sector emissions : 13 539 • Transport share in total : 30.5%	Not provided	Not provided	Not provided
Romania	• Road emissions : 6 832.2 • Sector emissions : -- • Transport share in total : -- Base year = 1993	• Road emissions to 2000 : + 48% to 2010 : +150%	• Road emissions to 2000 : + 48% to 2010 : +150% (same as Reference Case)	• Road emissions to 2000 : + 26% to 2010 : +105%
	Average change in annual transport emissions (Road Transport only) 1993-95 : +7.5%			
Russian Federation	• Road emissions : 146 900 • Sector emissions : 234 600 • Transport share in total : 9.6%	Not provided	Not provided	Not provided
	Average change in annual transport emissions 1990-1995 : -9.4%			
	Notes : The Ministry of Transport notes that reliable projections for 2000, 2010 have not yet been made because of the prevailing economic instability in Russia. The Ministry plans to develop projections in the near future.			
Slovak Republic	• Road emissions : 4 500.8 • Sector emissions : 5 296.2 • Transport share in total : 10%	• Road emissions: to 2000 : + 3% to 2005 : + 9% • Sector emissions to 2000 : + 2% to 2005 : + 9% • Transport share in total: 2000 : -- 2005 : --	• Road emissions: to 2000 : - 4% to 2005 : - 3% • Sector emissions to 2000 : - 5% to 2005 : - 2% • Transport share in total : 2000 : 11% 2005 : 11%	• Road emissions: to 2000 : - 8% to 2005 : - 9% • Sector emissions to 2000 : - 8% to 2005 : - 7% • Transport share in total: 2000 : -- 2005 : --
	Average change in annual transport emissions 1990-94 : -4.4%			
	Notes : The Slovak Ministry evaluates the quality of the data provided as grade C on a five-grade scale : of an approximate nature but sufficiently well-estimated to be considered representative.			
Slovenia	• Road emissions : 2 947 • Sector emissions : 3 192 • Transport share in total : 23.4%	Not provided	Not provided	Not provided
	Average change in annual emissions 1990-95 : +7.2%			
Spain	• Road emissions : 48 706 • Sector emissions : 74 332 • Transport share in total : --	Not provided	Not provided	Not provided

	BASE YEAR 1990 unless otherwise indicated (1000 tonnes)	**REFERENCE CASE** Projections 2000, 2010 (no additional measures since base year)	**STATUS QUO** Projections 2000, 2010 (with only measures implemented since base year)	**FUTURE** Projections 2000, 2010 (with measures implemented or planned)

Sweden	• Road emissions :
	  16 100
	• Sector emissions :
	  20 500
	• Transport share in total :
	  34.2%
		(Approximately the same figures as Status Quo case, according to the Ministry)	• Road emissions:
			  to 2000 : + 8%
			  to 2010 : +17%
			• Sector emissions
			  to 2000 : + 6%
			  to 2010 : + 7%
			• Transport share in total:
			  2000 : 34%
			  2010 : --
				• Road emissions:
				  to 2000 : +2%
				  to 2010 : - 9%
				• Sector emissions
				  to 2000 : +2%
				  to 2010 : - 7%
				• Transport share in total:
				  2000 : --
				  2010 : --

Switzerland	• Road emissions :
	  12 620
	• Sector emissions :
	  14 770
	• Transport share in total :
	  33.4%
		Data for Road, Sector and Share not provided	• Road emissions:
			  to 2000 : + 10%
			  to 2010 : + 22%
			• Sector emissions
			  to 2000 : + 11%
			  to 2010 : + 24%
			• Transport share in total:
			  2000 : 36.4%
			  2010 : 39.3%
				Data for Road, Sector and Share not provided

	Average change in annual transport emissions 1990-95 :+0.7%			

United Kingdom	• Road emissions :
	  109 691
	• Sector emissions :
	  119 744
	• Transport share in total :
	  21%
		• Road emissions:
		  to 2000 : +27%
		  to 2010 : +49%
		• Sector emissions
		  to 2000 : +26%
		  to 2010 : +47%
		• Transport share in total:
		  2000 : 26%
		  2010 : 28%
			Not provided	• Road emissions:
				  to 2000 : +14%
				  to 2010 : +29%
				• Sector emissions
				  to 2000 : +15%
				  to 2010 : +29%
				• Transport share in total:
				  2000 : 25%
				  2010 : 26%

	Average change in annual transport emissions 1990-94 : +0.55%			
	Notes : Emissions presented by UNECE source category. Projections are based on Energy Paper 65 model runs with (Future Measures) and without (Reference Case) Climate Change Programme. Figures are subject to increasing uncertainty into the future. Data provided to sector level; sub-sector data not readily available in consistent format -- further disaggregation for road sector will be available when recently-developed vehicle market model is validated.			

United States	• Road emissions :
	  1 165.0
	• Sector emissions :
	  1 583.6
	• Transport share in total :
	  33%
	(figures in million tonnes)
		• Road emissions:
		  to 2000 : +16%
		  to 2010 : +27%
		• Sector emissions
		  to 2000 : +17%
		  to 2010 : +31%
		• Transport share in total :
		  2000 : 34%
		  2010 : 35%
			• Road emissions:
			  to 2000 : +13 %
			  to 2010 : +23%
			• Sector emissions
			  to 2000 : +15%
			  to 2010 : +27%
			• Transport share in total :
			  2000 : 35%
			  2010 : 36%
				Not provided

	Notes : Projections performed on IDEAS Model Run AHMBW7 by Department of Energy.			

4. POLICIES AND MEASURES TO LIMIT CO_2 FROM TRANSPORT

In the second part of the questionnaire, ECMT countries were asked to provide information on policies and measures in place or envisaged to limit CO_2 from transport. Section 4.1 describes and briefly analyses some of the principal policy tools that can be used to target CO_2 from transport. Section 4.2 details the specific policies and measures that were cited as in use or under consideration in ECMT countries based on the information provided in the responses to the questionnaire. Section 4.3 then draws general conclusions about the types and frequency of policies cited by ECMT countries.

4.1 Description of policies and measures targeting CO_2 from transport

This section describes the principal policy options that have been studied and applied to limit transport's impact on climate change. There is considerable debate on the effectiveness and feasibility of these measures. The objective of this chapter is to provide an overview of these areas of policy development, highlighting their main advantages and disadvantages (discussion is by no means meant to be comprehensive), in order to set the context for the Section 4.2 description of policies and measures in use or envisaged in ECMT countries.

Responses to the 1996 questionnaire clearly show that governments recognise the need for changes in the way transport systems are organised, operated and used. In recent years, ECMT transport ministers supported policy actions to counter rising car dependency and encourage sustainable development of transportation systems. However, two years after the preparation of the ECMT Interim Report on Transport and the Greenhouse Effect, it appears clear that stronger political will is needed along with continued government-industry co-operation in order to carry policy ideas forward to implementation.

Policy packages exist that have potential to stabilise and reduce transport-related CO_2 emissions. In addition to government-driven regulatory measures or economic instruments, CO_2 abatement packages can involve voluntary initiatives on the part of industry. Packages of policies will differ from country to country, as each system has its own specific economic and regulatory climate in which policy development must take place. Whatever, the policies decided upon, if targets for beyond 2000 -- currently being established in the FCCC process -- are to be met, actions in most countries will need to be more effective than in the past.

Some measures are clearly politically and socially difficult to implement, especially where they concern setting fuel economy parameters for the vehicle industry, for example, or using fuel price increases to incite private car users to leave their vehicles behind and opt for public transport or other transport means where possible. Indeed, the economic and trade effects of certain measures are of considerable concern to industry, and demand for private transport remains defiantly high in the face of increasing fuel taxes. If the FCCC objectives are to be attained, however, the government-industry dialogue must continue to seek socially and politically acceptable ways to reduce transport-related CO_2.

The policy categories most often targeted for CO_2 emissions abatement include measures to:

-- improve the organisation and operation of the transport system
-- increase vehicle fuel efficiency
-- influence driver behaviour
-- develop the potential for alternative fuels

4.1.1 *Organisation and operation of the transport system*

Better traffic management, improved public transport and integrated land-use and transport planning are key factors in the long-term, environmentally sound development of transport systems. They can also contribute to CO_2 abatement -- if somewhat more modestly than, for example, vehicle fuel efficiency or fuel taxes.

Better traffic management

Better traffic management involves, *inter alia*, reducing the number of vehicles on the road, more effective parking policies and use of available traffic management technology.

-- Schemes for reducing the number of vehicles on the road

Car-sharing

Encouraging individuals and companies to seek opportunities for vehicle sharing has been shown in certain areas to be effective in reducing the number of cars on the road and thus improving traffic, especially in rush hours. Indeed, the average occupancy of cars is less than 1.5 in most cities (ECMT/OECD 1995). Attributed with responsibility for low car occupancy are higher vehicle ownership rates and income levels, greater numbers of women in the work force and smaller households. Low vehicle occupancy means increased vehicle travel, energy consumption and emissions. (US DOE, 1996).

Car-sharing schemes have mostly been implemented in congested areas, where benefits in terms of reduced vehicle numbers are most evident. As with a number of other traffic management actions, car-sharing's effect on CO_2 emissions can be diminished by a latent surge in demand, spurred by increased capacity (US DOE, 1996). A number of experiments have been conducted on car pooling in recent years -- especially in the United States -- with various results: for individuals, the time and effort required to organise such schemes seem to outweigh the potential benefits; they may be more effective as a traffic management tool when employers are charged with the responsibility of organising and handling the logistics (ECMT/OECD 1995). A 1988 scheme in California, which involved a regulation requiring employers to implement trip reduction programmes, brought about a modest increase in vehicle sharing from 14.1 per cent to 19.9 per cent, and a 5.2 per cent drop in the percentage of employees driving alone to work within the first year of the programme (US DOE, 1996). A further study tracked evolution in commuting behaviour in a representative number of companies over the following two years: the number of individuals driving alone dropped by about 6 per cent; car sharing increased by 40 per cent, while public transit use and other methods such as walking and biking decreased by 4.3 and 6.9 per cent respectively. It is important to note that commuting to and from work represented less than one-third of travel by the average US household in 1990 (1994 Cambridge Systematics study cited in US DOE, 1996).

This experience perhaps demonstrates that targeted policies to encourage both employer and private household-instigated vehicle sharing may lead to better results. Combining designated highway lanes for high-occupancy vehicles (HOV) with road pricing based on occupancy numbers may facilitate the success of car-sharing programmes.

Telecommuting

Promoting opportunities for telecommuting can also reduce the volume of commuter traffic, particularly at peak flow times. With significant advances in telecommunications in recent years (e.g. fax machines, electronic mail and data base management systems) workers have more flexibility as to their work place. Growing numbers of employees -- notably again in the United States -- are successfully working from their homes via communication links to their companies.

In 1992, a survey in the United States found that 4.2 million employees -- 3.3 percent of the workforce -- were telecommuting, a 27 percent increase from the previous year. A US Department of Transportation Study found that the number of telecommuters may grow to 10 million by 2000; more than 30 million by 2010 and up to 50 million in 2020 (US DOT, 1993).

As with car-sharing schemes, the effects of telecommuting on CO_2 abatement are estimated to be relatively modest, however. According to the US Department of Energy, additional commuters are expected to take to the roads in order to benefit from improved rush-hour speeds. In addition, the savings in commuting costs is expected to bring about increased dispersion in urban settlement. Together, these countervailing effects are expected to diminish by one half the positive impact of increased telecommuting. Nevertheless, the DOE report states that the projected increases in telecommuting should among others:

- reduce wasted commuter time by 160 million hours per year;
- avoid approximately $15 billion in urban highway and road construction spending;
- save 1.5 billion gallons of fuel and 10 million tonnes of CO_2.

-- More-efficient parking policies

Policies that encourage sound management of parking capacity can be instrumental in reducing congestion and its effect on the environment. Parking charges are the most widely used measure for restraining traffic in OECD countries.

Parking control measures should be carefully designed, however, as undesired effects can result. The balance of supply and demand is a delicate one: increasing capacity risks encouraging higher demand (more congestion). A 1992 study conducted by the Netherlands Agency for Energy and the Environment for the International Energy Agency found that people making short trips are more likely to be dissuaded by difficulty finding places to park than those making longer trips. The study found that when traffic from shorter trips is displaced, parking capacity is created for longer trips (IEA/OECD, 1993). The traffic-restraint objectives may therefore be lost.

In the United States, an innovative system known as "parking cash out" is being used in California and is under discussion on a federal level to modify company-subsidised parking for employees. Under the parking cash-out mechanism, employers are required to offer all employees the option of taking a tax-free allowance of equal value to the parking subsidy, which can be used for a variety of transport options including public transport use and car pools. In this way, the effect of the subsidy -- which is, in essence, an incentive for employees to take their private cars to work -- is reduced. The idea is that parking cash-out will "level the playing field" among transport modes for travel to work. The system is mandatory in California, but is being examined more as a potential voluntary action on a federal level. The numbers of employees who drive alone to work is estimated to drop 70 per cent to 55 per cent as a result of this measure (ECMT/OECD 1995).

-- Use of available technology for traffic management

A variety of technological tools are becoming available that are expected to help alleviate congestion and its effects on local and global environment; in particular, telematic motorist information systems and improved traffic-signalling systems. Timely, accurate information disseminated via telematics to home or work computers can help motorists plan their trips in the most efficient way -- noting available parking sites and alternative travel routes. In addition, integrated, real-time computer programmes that use congestion and incident detector systems to optimise traffic signal timing and give priority to on-road public transport vehicles can benefit traffic management. Considerable resources have been devoted by, among others, Japan, the United States and the European Union to the research and development of information technologies that are applicable to road transport. In the next ten years, these developments are expected to significantly improve the information available to traffic managers and transport users, thereby leading to more efficient driving and greater fuel economy. (ECMT/OECD 1995).

Access control to the city centres of Barcelona and Bologna has recently been tested using automatic vehicle identification based on digital images and licence plate reading technology. Among the results of the test for the controlled "special events" zone in Barcelona were:

- an average 18 per cent reduction in travel time inside the zone;
- a 33 per cent decrease in entry volumes to the area;
- a 55 per cent increase in public transport trips city-wide, a favourable modal shift effect.

(EC- DG VII/DG XIII, 1996).

As with other congestion mitigation measures, improved traffic management through use of advanced information technologies may actually lead to greater capacity and subsequently higher travel demand, energy use and emissions (NRC, 1995, cited in US DOE, 1996). Capping growth in private vehicle use via fuel and possibly road and congestion pricing mechanisms might, however, help impede the countervailing effects of increased demand.

Road pricing

Advanced technology solutions for transport systems are particularly interesting due to their ability to facilitate road pricing. Accurate tracking of vehicle road use (location, time of day) is necessary for sound road pricing schemes.

Road pricing involves a system whereby the number of kilometres driven on each road type is electronically recorded, with tariffs per kilometre set on road type, vehicle type and degree of congestion. Payment can be handled in the form of a prepaid card, from which the correct amount of fee is debited. The revenues are then be allocated to the different owners of the infrastructure (e.g. municipalities).

Further examination of both the technical and administrative aspects of road pricing will be necessary before general electronic road pricing is ready for implementation. Road pricing in urban areas may be a less-distant option. Austria is planning to adopt a road pricing scheme as part of its Five Points Programme. The Netherlands is looking into the introduction of road pricing after 2000. Sweden is also examining possibilities for road pricing. Individual privacy issues may surface in road pricing schemes, as noted by Wachs (Wachs, M., 1994, cited in US DOE 1996); however options such as automatic payment schemes that assure driver anonymity may assuage the privacy concerns.

Technology to facilitate and improve traffic management is only now becoming available, so the verdict is yet out on the extent to which it will prove to be significant in alleviating congestion. If social and other obstacles can be hurdled, however, indications are that it may be a key to mitigating the congestion problem.

Improved public transport systems

Improving the public transport system alone will most likely not provide sufficient incentive for drivers to relinquish their cars for mass transit. However, public transport development is an essential part of a sustainable transport policy package. Assuring availability of reliable, efficient public transportation can minimise the need for travel by car. High-quality public transport may help to retain jobs and commercial activities in urban centres, reducing the need for city residents to travel long distances. Without such public transport systems, modal switching, a key element in sustainable transport strategies, is unlikely. If better public transport service is not accompanied by a corresponding reduction in car use however, traffic congestion and emissions will most likely not improve.

ECMT countries, as shown in the responses to the 1996 questionnaire, consider improvements to their public transport systems as an important part of their plans to limit CO_2 from transport. This is especially evident in, though by no means exclusive to, many Central and Eastern European countries, the public transport systems of which are in need of re-organisation and renovation. Measures reported by ECMT countries in their responses to the 1996 survey reveal support for public transport primarily via use of subsidies or other forms of government investment. While countries believe that subsidies and investment packages are essential to assuring a high quality transport system and services, they may not be sufficient to attract and keep passengers, however: competitiveness of public transport also requires better response to client (passenger) needs, cost recovery and overall system efficiency. Information campaigns encouraging drivers to opt for public transport where possible may also be instrumental in increasing public transit use.

Integrated land-use and transport planning

Key to long-term efficiency of transportation systems is a meaningful and sustained dialogue between land-use planners and those in charge of designing and implementing transport system development. OECD countries have had difficulty coming to terms with this aspect, evidenced by severe congestion in urban centres and often-chaotic urban sprawl. Countries are, however, beginning to integrate this approach into their planning strategies: a number of ECMT countries cited examples in their survey responses of actions planned or envisaged to move toward a combined approach to land and transport planning. By considering where people live and where activities take place, land-use policies can have a positive affect on what kinds of trips are made, the distance travelled and the mode of transport used.

The link between land-use policies and transport-related greenhouse gas emissions is less direct than that between traffic management policies and these emissions; nevertheless the importance of land-use policies appears certain: effective, long-term land-use planning can influence transport-related emissions levels by minimising the need to travel -- thereby reducing the number of cars on the road -- and maximising the capacity of public transport.

Land-use planning can limit the need to travel in a number of ways; notably by minimising the distances between places of residence, employment, and commercial or other activities. This is particularly pertinent given that traffic growth in OECD countries over the last 50 years can be attributed more to longer distances per trip per person per day (pppd) than to a greater number of trips pppd. Land-use planning can also enhance the benefits public transport provides in terms of fewer cars on the road. This involves ensuring that trip-attracting activities are concentrated in locations that are easily accessible by public transport and targeting areas well-served by public transport for new residential, commercial and business developments (ECMT/OECD, 1995).

Several problems in evaluating the impact of integrated transport and land-use planning on travel demand (and therefore on vehicle kilometres travelled and subsequent CO_2 emissions) bear mention: first, further empirical studies are needed in order to better understand the relationship of the two. In addition, the ways to go about implementing, as well as the obstacles to integrated land-use and transportation planning must be more clearly understood before the efficacy of these policies on CO_2 abatement can be more precisely evaluated (US DOE, 1996).

Better traffic management, improved public transport and integrated land-use and transport planning enhance the efficiency of the overall transportation system by facilitating reduced car use, fuel consumption, and vehicle emissions. In this way, they are important in assuring the sustainable development of transport systems. According to studies to date, the direct effect of these policies on CO_2 abatement appears, however, to be somewhat limited. The next two sections -- vehicle fuel efficiency and alternative fuels -- examine two vehicle-related aspects that more-directly influence the emissions of greenhouse gas from transport.

4.1.2 *Vehicle fuel efficiency and fuel pricing*

Vehicle fuel efficiency is a determining factor in emissions of CO_2 and other gases. A 10 per cent improvement in vehicle fuel efficiency at present can result in a 7 to 8 per cent reduction in greenhouse gas emissions. Fuel economy can be significantly improved by reducing the size, weight and power-to-weight ratios of vehicles, as well as by increasing the efficiency of the drivetrain (engine, transmission and axles) and decreasing the work required to move the vehicle (tractive effort) including raising tyre pressure (ECMT, 1993a). Use of currently available technologies and better vehicle and engine design may be capable of bringing about significant reduction in specific CO_2 emissions; however, beyond a certain level, decreased vehicle performance has to be accepted. Best available technology can make a difference here, but it must be kept in mind that to date, there is no known "technological fix" for CO_2.

Apart from pricing, the principal mechanisms for encouraging improved new vehicle fuel economy are standards, targets, feebates and voluntary initiatives on the part of industry. There is considerable discussion surrounding the effectiveness -- particularly cost-effectiveness -- of these measures. Indeed, they offer a variety of beneficial effects, but can also carry with them a number of uncertainties. For example, while they can provide incentives to vehicle manufacturers to develop new technologies and re-direct research efforts from improvement of performance, comfort and safety into better fuel economy, the long-term nature of their effects and the results of the research are difficult to determine. The measures can also encourage manufacturers to market and consumers to purchase more energy-efficient/small vehicles. These effects, however, will be felt only as long as the measures are in place. Further, measures to improve vehicle fuel economy can lead to what is known as the "rebound effect": an increase in vehicle efficiency will produce a drop in overall costs of driving, and therefore a subsequent increase in vehicle kilometres travelled and a potential surge in vehicle purchases (Michaelis, 1996), counter to the overall emissions reduction effects desired.

Cars became more fuel-efficient throughout the OECD from 1975 to the early 1990s, largely due to progress in vehicle and engine technology. According to the European Automobile Manufacturers Association (ACEA), the average fuel consumption of the new car fleet in Europe dropped from 8.3 litres per 100 km in 1980 to 7.1 litres per 100 km in 1995 -- an almost 15 per cent reduction. In most OECD countries, on-the-road average car fuel consumption is approximately 8-12 litres per 100 km, and cars with a fuel consumption of 2.5-3 litres per 100 km have been produced in prototype form (IEA/OECD, 1993). However, trends show that consumers continue to seek larger, more powerful cars, making it more difficult for manufacturers to produce more fuel-economic models. For this reason, a combination of stronger regulatory and economic instruments may be necessary to promote fuel efficiency efforts on the part of both vehicle manufacturers and consumers.

Standards for new-vehicle fuel efficiency

Establishing standards for fuel efficiency can have positive effects on fuel economy in cars in some cases, and these benefits may be enhanced when standards are implemented along with fuel pricing policies or vehicle tax/rebate schemes designed to encourage the purchase of more fuel-efficient cars. In their responses to the 1996 survey, the Czech Republic and Russia cited plans in place or envisaged to set vehicle fuel efficiency standards; Japan noted tightened targets for vehicle fuel efficiency; and Switzerland reported a new 1996 fuel economy target.

The United States is at present the only ECMT country that has a mandatory vehicle fuel efficiency standard in place. The new car fleet's efficiency virtually doubled in the United States from 1974 to 1985 with the establishment in 1978 of a Corporate Average Fuel Efficiency (CAFE) standard at 18 mpg[6] and higher fuel prices. CAFE standards were subsequently increased to 27.5 mpg[7] in 1985, however, low fuel prices in the United States since then have not encouraged consumers to purchase these more-efficient vehicles (Nadis and MacKenzie 1993). Indeed, the only period in which the trend toward larger, more-powerful cars was stopped was the mid-1970s, when fuel prices were high. It should also be noted that body and engine downsizing at the time coincided with availability of improved engine and body technology. At present, there appears to be no downsizing effect left from CAFE; the trend toward larger and faster cars continues to prevail. CAFE-type standards can play a role in guiding industry developments; they appear to be most effective, however, when implemented in tandem with robust pricing measures to influence consumer behaviour.

The debate continues on this issue. Analysis from the US Department of Energy suggests that if fuel standards are implemented in an appropriate time frame and set at levels that: are able to be met by technological changes, do not impact vehicle utility, and provide fuel savings equal to the increased cost of fuel-efficiency technology, then they may be effective and well accepted by consumers (US DOE, 1996).

Vehicle tax/rebate system

Another policy approach to new-vehicle fuel efficiency involves assessing fees on vehicles that do not meet fuel-economy standards and offering rebates on those that do. Under the so-called "feebate" system, new, fuel-inefficient vehicles are taxed according to their litre/100 km or mpg rates. The revenues generated by such taxation are pooled in a fund, which is then used to provide rebates on the purchase of more fuel-efficient vehicles that meet CAFE-type standards. A feebate plan is currently in effect in Ontario, Canada, whereby a fuel-consumption-linked tax of C$ 75 to $ 4 400 is assessed on all cars surpassing a 6 litre/100 km standard. Purchasers of vehicles with a fuel-efficiency rating under this standard can benefit from a rebate of up to C$ 100 (Canada, 1994).

Whereas with CAFE-type standards, the size of the incentive to improve fuel economy differs among manufacturers, and therefore carries implications for competitive fairness and trade, feebates offer all manufacturers the same incentive in terms of the marginal value per vehicle model of reducing fuel consumption. In this way, feebates may be preferable to standards. In order to be effective, however, feebates need large-enough car markets for car design to be influenced; vehicle markets in individual European countries or North American provinces and states may not be of sufficient size to bring about conclusive fuel economy benefits. Markets of relative consequence, however, such as those in the European Union, Japan or the United States may be adequate (Michaelis, 1996).

Programmes to remove older vehicles from the fleet

Ridding the vehicle fleet of old, fuel-inefficient, polluting vehicles can also help to improve overall fleet fuel economy. The US Department of Energy has estimated that removing 2 million old-model vehicles from the road could result in an annual savings of 150 to 230 million gallons of fuel. (Nadis and MacKenzie, 1993). ECMT countries reported a number of car-scrapping programmes in their responses to the questionnaire. In 1994-95, the French government offered FF 5 000 toward the purchase of new cars to car owners willing to retire their old models of at least 10 years of age, an initiative presented primarily as an economic stimulus to the sagging automobile industry in France ; 400 000 vehicles were withdrawn from circulation due to this measure (France, 1995). A similar programme came to an end in September 1996. Spain has launched two fleet renewal plans since 1994 as well. In its 1993 fleet renewal programme, Hungary offered Forint 30 000 in public transport tickets in exchange for old Trabant vehicles ; 10 000 vehicles were traded in as a result of this initiative.

Car-scrapping programmes such as these can have positive effects on overall air quality, reducing potential emissions from older, polluting cars. And given that more-recent car models are more fuel-efficient, the overall fuel economy of the fleet can improve from these initiatives. The benefits may, however, be less substantial than those achieved for criteria pollutant emissions. Car scrapping schemes may only be beneficial when targeted at very old models and less-developed vehicle fleets. In practice, these schemes may even be counter-productive on a full-cycle basis as concerns CO_2 emissions due to manufacturing losses, increased mileage, and the purchase of new models from the same class that do not offer real gains in fuel economy. Analysis cited by the US Department of Energy shows that these programmes can be cost-effective and provide fuel savings and CO_2 emissions benefits if : applied to the highest emission vehicles only; prices paid for the older vehicles are appropriate; only registered vehicles are purchased (US DOE, 1996). These benefits may not, however apply in the same way to the European vehicle market, the overall configuration of which comprises smaller cars relative to that of the United States.

Voluntary action

Some argue that regulatory measures such as fuel efficiency standards and feebates are undesirable, given the costs to manufacturers of developing and applying new technology and to the environment due to the propensity of drivers to increase travel in response to lower costs of driving. Industry representatives most often favour voluntary initiatives to improve vehicle fuel efficiency. Voluntary actions may provide a mechanism to minimise dissent between government and industry groups (Michaelis, 1996).

In the Annex 1 national communications provided so far to the FCCC, measures addressing transport-related greenhouse gas emissions focus heavily on regulations and economic instruments such as fuel taxes. Voluntary initiatives remain relatively few in number. In a recent International Energy Agency survey of voluntary actions in OECD countries, less than 10 per cent of the 200 voluntary initiatives reported were in the transport area. A trend towards voluntary actions was, however, noted in the report (Solsbery and Wiederkehr 1995).

On an international level, several voluntary initiatives bear mention. In June 1995, the Council of the ECMT, comprised of member-country transport ministers, and the vehicle manufacturing industry, represented by ACEA and the International Organization of Motor Vehicle Manufacturers (OICA), signed a joint declaration on the reduction of CO_2 emissions from passenger cars, the objectives of which are to "substantially and continuously reduce the fuel consumption of new cars sold in ECMT countries", and "to manage vehicle use so as to achieve tangible and steady reductions in their total carbon dioxide emissions" (ECMT, 1995). The parties in the declaration agree to study opportunities for: a car-labelling system; improved co-ordination of research and development; identifying criteria for information technology in vehicles; and developing information/education campaigns aimed at vehicle users, dealers and importers. As part of the accord, ECMT is currently monitoring both specific fuel consumption and CO_2 emissions of new vehicles registered each year, as well as the policies and measures introduced nationally to reduce emissions (covered in this report).

The Council of the European Union has recently mandated the European Commission to open negotiations with European vehicle manufacturers and importers for average vehicle fuel consumption targets of 5 litres per 100 km for new petrol cars and 4.5 litres per 100 km for new diesel cars. These objectives are the equivalent of 120g CO_2 per km -- a 30% to 35% improvement -- between 1990 and 2005 (EU Council, 1996).

Voluntary action offers considerable potential to stimulate improvements in average new-vehicle fuel economy. In government-industry agreements, vehicle manufacturers agree to achieve fuel economy targets as a group ; therefore responsibility for meeting the targets is distributed among individual firms. Important steps are being taken to reach voluntary accords in ECMT countries; it remains to be seen how effective these agreements will be in improving overall vehicle fuel efficiency.

Fuel pricing

The role of fuel prices in CO_2 emissions abatement has also been the focus of much study and debate. Given that the objective is to reduce fuel consumption, be it by means of improved vehicle fuel economy or lower mileage driven, higher fuel prices can act as an incentive to produce and purchase more fuel-efficient cars, or simply drive less. They can also offset the potential countervailing cost effect of CAFE standards that in isolation can stimulate increased propensity to drive. Indeed, fuel taxes can be a powerful tool in influencing overall driver behaviour, thereby affecting traffic congestion and vehicle emissions.

From an economic standpoint, fuel taxes are often described as the most efficient means of reducing consumption of fuel. Studies in Germany and the Netherlands suggest that taxation policies -- in these cases involving substantial tax increases -- were most effective among a number of measures in reducing transport-related greenhouse gas emissions (ECMT 1993a). The question of the elasticities of consumer choices relative to different taxation levels in the short, medium and long terms is a highly debated topic.

The political feasibility of sizeable tax hikes is an issue, as well, given that substantial price hikes may lead to public resistance. One way to avoid this might be to attenuate the impact of higher prices through a gradual and progressive rise in real fuel prices. In this way, consumers can gradually adapt behaviour to the price changes (ECMT 1993a). Avoiding tax-related distortions by reducing taxation in other areas is also important to assure the overall efficiency of the market.

In their responses to the 1996 survey, numerous ECMT countries recorded tax schemes as providing incentives for more efficient fuel use and better vehicle fuel economy. In Austria, for example, a trend toward larger-engine vehicle capacities appears to have been halted for the first time in 1993 as a result of a 1992 standard fuel consumption tax and a 1993 engine-based tax. A one-off mineral oil tax increase has been considered as a means to limit growth trends in vehicle performance. Revenues from this tax increase are expected to amount to approximately ATS 5.6 billion per year, earmarked for the extension of local rail infrastructure (Austria, 1994). The United Kingdom decided in 1993 to increase fuel taxes by an average of 5 per cent per year above the inflation rate (ECMT/OECD, 1995). Road fuel demand fell in 1995 for the first time since 1991, suggesting that the road tax might be having an effect on demand (Cambridge Econometrics, 1996).

4.1.3   Driver behaviour

It is widely recognised that the way drivers use their vehicles can significantly affect vehicle fuel consumption and emissions. Research conducted for the Netherlands Agency for Energy and the Environment (NOVEM) has shown that differences in driving style can account for a variation of up to 50 per cent in vehicle fuel consumption among drivers using the same cars. Realistically achievable savings from driver training and the use of simple on-board instruments -- econometers and cruise controls -- is probably in the order of 15 per cent. Factors leading to improved fuel consumption include:

- avoiding excessive idling of the engine;
- driving smoothly (avoiding high revs);
- limiting high-speed driving (in general terms, fuel consumption and pollution significantly increase above 80 km/h, particularly above 100 km/h);
- maintaining adequate tyre pressure;
- eliminating sources of unnecessary drag.

Education and information are critical in helping drivers to learn to use their vehicles more fuel-efficiently. As concerns driver education, integrating fuel-efficiency into the curriculum of standard driver training courses can catch budding drivers before they have time to develop inefficient driving habits. Informing already-seasoned drivers on ways to modify their driving behaviour to save fuel can also be effective; it has been shown that if drivers are aware of the potential savings involved in driving more efficiently, they will often modify their driving habits accordingly. Training of truck drivers in fuel-efficient driving techniques can lead to significant savings for trucking companies when combined with appropriate scheduling. And commercial fleet operators, who tend to take fuller account of costs in their management decisions than drivers of private vehicles, can also benefit greatly from training in more fuel-efficient driving. The key to the success of measures to improve driver behaviour is information. Campaigns to increase public awareness of the importance of better driving in fuel efficiency are the basis for changing the habits of a significant number of drivers and improving overall fuel consumption.

Econometers and other in-car technologies, which inform drivers as they travel of how their vehicle is using fuel, can reinforce information and education programmes. Possible incentives to develop the market for these instruments could be envisaged. Better enforcement of existing regulations, for example, -- particularly as concerns speed limits -- is essential.

Numerous ECMT countries cited tighter enforcement of speed limits, driver education and information campaigns designed to influence driver behaviour as measures to improve fuel efficiency and reduce emissions. As an important part of its greenhouse gas emissions control strategy, the Netherlands, for example, has set a target of 10 per cent fuel savings and CO_2 emissions reductions to be achieved by improved driving behaviour and lower driving speeds, and has developed an action programme on fuel-efficient driver behaviour and vehicle purchasing practices. Canada has also put into place several comprehensive information and education programmes targeting specific sectors and fleets to improve awareness of vehicle fuel economy and fuel efficient driving.

4.1.4 Alternative fuels and electric vehicles

Government and industry have for many years been seeking ways to alleviate transport's heavy dependence on fossil fuels. For reasons of both supply security and the environment, considerable resources have been devoted to the development of alternative fuel options. Whereas vehicle fuel-efficiency measures and fuel taxes described above can be envisaged in a relatively short time frame, the widespread use of non-petroleum sources of energy for transportation, however, remains a more-distant solution, despite substantial technological advances and limited successes in recent years. Until the cost of switching to other fuels becomes more competitive, and the operation of alternative-energy vehicles less compromising on performance and range, it is unlikely that vehicle users will make the switch in significant numbers.

Measures to address use of alternative fuels include: increased taxation of petrol and/or diesel fuels, financial incentives to purchase alternative fuels (including lower taxation than on conventional fuels), financial support for the purchase of electric/natural gas vehicles, research and development. Following is a brief look at some of the alternative fuels under study and examples of ECMT country initiatives to address their use.

Liquid petroleum gas (LPG) includes butane and propane and is derived from the refining of crude oil and natural gas. Its principal advantage over petrol is its high energy content and relative low cost. On an overall fuel-cycle basis, LPG -powered vehicles may be able to achieve a 20 per cent reduction in greenhouse gases compared to petrol. **Compressed natural gas (CNG)** is primarily methane, so an important factor in the effectiveness of CNG as a favourable fuel in the fight against greenhouse gases is minimisation of methane emissions. As long as these emissions can be controlled, its greenhouse gas reduction potential compared to petrol is approximately 20 per cent (primarily CO_2 emissions). (ECMT, 1993a). One disadvantage of CNG is that fuel tanks must be stronger and larger than those for petrol to assure safe storage of the fuel. As a result, vehicle weight and range is affected. The number of CNG vehicles world-wide has grown by an estimated 50 per cent over the last five years to more than 1 million vehicles. LPG vehicles also number more than 1 million around the world. The Japanese government has been subsidising the purchase of CNG-fuelled vehicles and fuelling facilities since 1994 (Japan 1994). Taxes have been reduced for these vehicles, as well as for electric, hybrid and methanol-fuelled vehicles. Public service administrations have also been granted funds for the use of low-emission cars. France has introduced a tax exemption effective 1 January 1997 for use of CNG, LPG and electric cars. Italy and the Netherlands among others have also implemented strong tax incentives for use of LPG and have sizeable numbers of LPG vehicles in their fleets. And the city of Vienna has been operating LPG-fuelled buses since 1976. In their response to the survey, the Slovak Republic reported a 50 per cent reduction in road tax for LPG and CNG vehicles.

Liquid biofuels: Canada and Belgium have both taken steps to exempt all or portions of biofuel mixes from taxation for the last four years. The Canadian government has also established an Action Plan to concentrate on research and development of ethanol production technologies along with assessments of economic and environmental impact of ethanol use in view of developing a commercially viable ethanol industry. (Canada, 1994 and Belgium, 1994). As a part of its recent air pollution legislation, the French government announced its intention to render obligatory by 2000 the use of biofuels such as ethanol in fuels.

Electric vehicles: Emissions associated with electric vehicles derive not from the vehicle, but from the source of energy. If the electricity generated for the vehicle is from coal-fired power plants, CO_2 emissions would be more than that of petrol vehicles. However, if the electricity is from hydro- or nuclear-powered plants, emissions would be significantly less relative to petrol vehicles (Nadis and MacKenzie 1993).

Many initiatives have already been taken to explore and develop electric vehicles as a viable option to petrol cars. However, to date, their performance and range capacity -- largely due to battery technology -- is limited compared to petrol vehicles, and their cost remains high. At present, they are being integrated to a limited extent into urban public transport and public service fleets, and are being considered for households as a second vehicle for exclusively urban use. In the United States, development of a viable electric vehicle option has been largely driven by stringent clean air standards that were adopted in 1990 in the state of California and subsequently adopted by 12 other states and

Washington, D.C. The 1990 regulations required 2 per cent of the cars sold in the state to be emission-free by 1998 and 5 per cent starting in 2001. In late 1995, however, the California Air Resources Board announced its intention to relax these requirements based on industry scepticism that necessary technology would be available by the target dates.

Hybrid vehicles combine the internal combustion engine with a second power source to optimise operational conditions. Since the principal power source is fuel, on-board storage of energy represents less of a problem than with electric vehicles. In this way, hybrids may provide a transition solution from exclusively internal combustion vehicles to electric vehicles. The alternative power source considered to be the most viable in terms of versatility, efficiency and size is an electrical system comprised of an energy storage mechanism, such as batteries or flywheels. Their reliance on fossil fuels as a primary power source means, however, that hybrids will not achieve zero-emission status. [EC (Joint Research Centre), June 1996].

4.1.5   The importance of packages

In the effort to reduce transport's contribution to climate change, the introduction of market-based incentives accompanied by complementary regulatory interventions appears to be the best approach. Technical and economic measures must accompany research and development initiatives and improvements in overall transportation planning and operation. Further, education and information programmes must be developed to win car users over to the importance of these actions and provide the information necessary for markets to function effectively; encouraging modifications in driver behaviour is a critical part of the success of short-, medium- and long-term greenhouse gas abatement from transport. Finding the right mix and strength of policies is now the challenge to ECMT countries. It appears clear that more decided steps need to be taken by government, industry and car users together in order to meet commitments made at Rio. There is no one recommended package of policies; individual governments must examine options and choose the right policy mix, acting where appropriate or necessary in common initiative with other countries.

4.2   Policy initiatives in ECMT countries

As noted earlier, recipients of the questionnaire were provided in Table 7 of the survey document with preliminary descriptions of their countries' policies and measures to limit CO_2 based on the National Communications to the FCCC. They were asked to modify/add to/update the information provided. All but a few countries completed the table. In their responses to the survey, ECMT countries were requested to classify policies and measures targeting CO_2 as one of the following:

- **R/G**   Regulation/Guidelines
- **EI**   Economic Instruments
- **VA**   Voluntary Agreement/Actions
- **IN**   Information initiatives/ **TR** -Training
- **R&D**   Research and Development

The status of the measures was then to be indicated as:

- **I**   Approved by Government and Implemented
- **A**   Agreed by Government -- not yet implemented
- **P**   Proposed in policy statement --agreement pending
- **E**   Envisaged as potential measure -- no official policy proposal made
- **O**   Other

Tables 4-8 house descriptions of policies and measures to target CO_2 from transport indicated by the ECMT countries which completed the questionnaire. Information on countries that did not fill in Table 7 of the questionnaire was taken primarily from the National Communications to the FCCC.

Please note that policies and measures have been included in the tables only if they were specific enough to classify in one of the categories indicated above. Initiatives such as "covering of external costs of transport" and "promotion of urban mass transport" were often listed that were too general to categorise in this type of matrix. The classification of measures given is by no means absolute; indeed, while some measures such as fuel taxes are clearly economic instruments, others could easily be included under several categories. The presentation here is meant to facilitate comparisons where appropriate and provide a general structure for analysis.

Table 4a. **ECONOMIC INSTRUMENTS**
Austria, Belgium, Canada, Czech Republic

Policies/Objectives cited	Austria	Belgium	Canada	Czech Republic
Fuel taxes to discourage excessive use of road vehicles	Mineral oil tax increase of Sch 0.50/litre petrol (I: since 1 January 1994).	Fiscal incentives to reduce road freight transport measures and render rail and water more attractive; support of combined transport; aid for interface structures (P). Tax increase on leaded, unleaded and diesel fuel (less on unleaded) (I).		
Fuel tax exemption/ reduction to promote use of certain fuels		Decrease in tax on bio-fuels used in pilot projects (I).		
Government subsidies/ other support to encourage use of public transport / inter-modal transport / modal shifting / discourage excessive road transport				Subsidisation of railway corridor development (I,A). Support to urban passenger transit (I).
Use of subsidies to encourage use of alternative-powered vehicles / new technologies			Financial incentive programme to promote natural gas as vehicle fuel and to encourage development of natural gas vehicle market: $500 for vehicle conversion; $1 000 towards purchase of new natural gas vehicle; $500 towards installation of vehicle refuelling appliance; $50 000 for each new refuelling station. (I: 33 000 natural gas vehicles in use and 200 public and private fuelling stations built).	Promotion of research and use of alternative fuelled vehicles via economic instruments (A).
Road taxes and road pricing	Part of Five Points Programme : Austria-wide toll system, involving in first stage -- motorways; second stage -- all trunk roads (A).			

Table 4a. **Economic Instruments** cont'd

Policies/Objectives cited	Austria	Belgium	Canada	Czech Republic
Fiscal measures to promote sale of more fuel-efficient vehicles	Vehicle sales tax based on fuel economy and vehicle price (I: 1 January 1992); engine-related insurance tax (I: 1 May 1993).		• 10 cent federal excise tax waived on alternative transportation fuels (I). • Elimination of federal excise tax of 8.5 cents/litre on the alcohol portion of petrol -- ethanol and petrol methanol -- fuel blends. (I: since 1992. Along with information and training actions, these fiscal measures have resulted in a 2% market share for these fuels in Canada).	
Fiscal advantages for public transport companies		Fuel tax reduction for regional public transport companies (I: 1992 law prolonged in 1993).		
Fiscal measures to encourage use of public transport for professional commuter traffic / reduce incentive for business travel		• Four measures restricting tax deductions on work-related vehicle expenses; car travel costs; private use of company cars (I: in effect since 1990, 1992 and 1993). • Fiscal incentives for employer subsidy of employee public transport passes and for employee use of public transport for travel to work (I).		
Tariff adjustment to encourage public transport use				Regulation of tariffs in favour of public passenger transport (I).
Actions to support combined transport	Government subsidies to combined transport under 1991 Master Transportation Concept (I).	Fiscal advantages for combined transport carriers (I).		Support for development of combined transport infrastructure (I). Fiscal advantages for combined transport carriers (I).

Table 4b. ECONOMIC INSTRUMENTS
Denmark, Finland, France, Germany

Policies/Objectives cited	Denmark	Finland	France	Germany
Carbon or CO_2 tax	CO_2 tax on all forms of energy use from non-renewable sources except natural gas and petrol (I: since December 1991).	Carbon and energy taxes in effect since 1 January 1990. Since 1994 covers all energy forms: fixed tax per tCO, and Mwh; on petrol means 0.13 FIM/l (1996) (I).		
Fuel taxes to discourage excessive use of road vehicles	Fiscal reform of 1993 providing for increases in fuel taxes; modifications in tax structure for goods vehicles and small lorries (I).	Fuel tax ~3.00 FIM/l (1.50 FIM/l diesel). (I: 1996/97 tendency is to increase the share of CO_2-related tax even though the overall tax level remains the same).	Progressive increase in minimum excise tax rates on fuel for HGV transport (I: Excise tax on diesel fuel is 31% higher than the minimum EU rate).	Increase in mineral oil tax in 1991. In 1994, mineral oil tax raised on leaded, un-leaded and diesel fuels (fuel tax hike less on diesel, but vehicle tax on diesel-engine vehicles raised in compensation) (I).
Fuel tax exemption/ reduction to promote use of certain fuels		The lower level of fuel tax on unleaded reformulated petrol and nonsulphur diesel has, during the 1990s, resulted in the share of these cleaner fuels growing to over 90% of consumption. (P)	Tax exemption for use of natural gas / electric / LPG-powered cars (A: Effective 1 January 1997).	

Table 4b. **Economic Instruments** cont'd

Policies/Objectives cited	Denmark	Finland	France	Germany
Government subsidies/ other support to encourage use of public transport / inter-modal transport / modal shifting / discourage excessive road transport	Fund set up to finance experiments with alternative public transport services outside regional centres (status not communicated). Strengthening of railway system via infrastructure investment as part of European high-speed rail network / promoting combined road / rail transport (status not specified).	- Reduction in transport subsidy granted to industries in developing areas (mostly situated in periphery) (I). - Subsidies for public transport. (Reduced sharply in recent years, however, because of recession) (I). - Investment in electrification of rail network (I). - Reduced VAT (6%) for public transport tickets (normal VAT is 22%, also on car purchase and petrol) (I). - Privatisation of railways in order to promote competitivity (I). - Development of high-speed passenger train connections (I).	- Investment in modern urban transport technology (I: From 1989 to 1993, national subsidies of FF 1.3 billion per year; State to maintain support of 5.5 billion per year 1994-1998). - Investment in infrastructure development for inter-modal transport (I: FF 250 million invested in conversion of major rail freight lines to wider gauge since 1985; new trading hubs to be introduced 1994-2000 in Bordeaux, Lille, Lyons, Marseilles by 2000 (partially I). - Development of high-speed TGV links to promote rail transport for inter-city travel [I: in 1993, 1 260 km of new rail lines in operation (1990-1993, FF 7.8 billion per year invested)]. By 2000, Mediterranean and eastern France TGV links will be in operation at an estimated investment of FF 55 billion (A).	- Federal subsidies for the development of public transport (suburban railways, subways, city trams, bus systems). (I: Since 1967, DM 50 billion has been spent for these purposes). - Federal government assistance for the construction and development of freight centres; to increase efficiency and promote co-operation in goods transport by bringing together companies that provide transport and related services for packaging and distribution of goods (I). - 1992 Federal Traffic Infrastructure Plan: Investment plan calling for DM 118 billion for railway network; DM 109 billion for long-distance motorways and DM 16 billion for waterways out of a total investment of DM 243 billion in construction and expansion of transport infrastructure 1991-2012 (I).
Use of subsidies for traffic management systems			- 1993: suburban traffic management systems introduced in Ile de France and Lyons conurbations -- to be fully operational by 1998. Capital investment of FF 1 200 million; operational costs FF 85 million per year (partially I).	DM 90 million spent in 1993 by the Federal Ministry of Transport on traffic-control measures for motorways: 70 traffic control systems now in operation. (I: Programme scheduled to be in effect through 1997, with 60 new systems planned).

Table 4b. **Economic Instruments** cont'd

Policies/Objectives cited	Denmark	Finland	France	Germany
Use of subsidies to encourage use of alternative-powered vehicles / new technologies	Subsidies to cover extra investment costs in purchase of energy-efficient and environmentally-friendly buses; trials with cleaner engines; alternative fuels and new technologies (status not specified).		Fund created 1 May 1991 to enable local authorities to purchase 1 000 electric vehicles (I: since 1992, purchase of 350 vehicles facilitated).	
Road taxes and road pricing			Basic road tax ("*vignette*", varies according to the "*département*", based on vehicle engine and transmission characteristics (I). Modification of *vignette* under study to better impact energy consumption by basing it on carbon content (E).	Since 1 January 1995, Germany in conjunction with the Benelux countries and Denmark has introduced a time-oriented motorway-use toll for trucks of at least 12t total permissible weight (I).
Fiscal measures to promote sale of more fuel-efficient vehicles		Taxation designed to promote sale of fuel-efficient cars (P).	- Vehicle registration tax = average FF 550 based on "fiscal horsepower", a function of engine size, power rating and fuel (petrol/diesel) (I). - Private firms purchasing electric cars and batteries are authorised to depreciate these over a 12-month period (I: since 1 December 1991).	1st phase: Conversion of existing motor-vehicle tax (targeting engine displacement) into emissions-oriented tax targeting commercial vehicles with a total permissible weight exceeding 3.5t. (I: in effect since 1 April 1994). 2nd phase: Conversion of existing motor-vehicle tax (targeting engine displacement) into emissions-oriented tax targeting light commercial vehicles, private cars and motorcycles (A: scheduled to be in effect as of 1 April 1997).
Vehicle fleet renewal			From February 1994 to 30 June 1995: FF 5 000 in cash paid for every vehicle over 10 years old scrapped when new vehicle purchased. (I: 400 000 vehicles withdrawn from circulation due to this measure). This incentive plan followed by similar one which ended 30 September 1996, offering FF 5 000 for small vehicles and FF 7 000 for medium-sized vehicles (I).	

Table 4b. **Economic Instruments** cont'd

Policies/Objectives cited	Denmark	Finland	France	Germany
Fiscal advantages for public transport companies			Reimbursement of excise tax on petroleum products (TIPP) for public transport (A: Effective 1 January 1997).	
Fiscal measures to encourage use of public transport for professional commuter traffic / reduce incentive for business travel	Reduction of tax reimbursement on diesel fuel in VAT-registered firms from all tax to DKr 1.10 per litre (I: in effect since July 1991).	Reduction in tax deduction of business travel expenses (I).		

Table 4c. ECONOMIC INSTRUMENTS
Hungary, Ireland, Italy, Japan

Policies/Objectives cited	Hungary	Ireland	Italy	Japan
Fuel taxes to discourage excessive use of road vehicles		Lower excise duties on diesel fuel than on petrol to provide tax advantage for ownership and use of diesel cars [I: super unleaded petrol increased by 3.1 pence per litre (VAT inclusive) effective 1 September 1996].	Increase in petrol/diesel tax (status not communicated).	
Government subsidies / other support to encourage use of public transport / inter-modal transport / modal shifting / discourage excessive road transport		Improved bus services outside Dublin area to encourage public bus use for inter-city travel; opening of new routes; upgrading existing routes by increasing frequency of service; modernising bus fleet. (I: substantial growth seen in long-distance travel by bus. Number of passenger journeys increased each year between 1987 and 1991). Improved urban and commuter public transport services in Dublin (I: introduction of electrified commuter rail service; development of diesel-based commuter rail services on existing mainline rail links; introduction of bus priority measures). State subsidies and capital investment in railways. (I,P: Direct state subsidies totalling over Ir£ 1 030 million between 1980 and 1992. Capital investment of almost Ir£ 250 million by railway company. Examples: improved track and signalling on mainline network; higher quality rolling stock. Further investment in plant and equipment planned in coming years, including up-grading of Dublin-Belfast line).	Promotion of modal switching away from road transport to reduce share of road in overall modal split (I: In effect since 1995). Development of pedestrian and bicycling areas (I: infrastructure development in effect since 1995). Promotion of rail use for freight transport (I: infrastructure development in effect since 1995). Development of high-speed passenger trains to reduce air traffic (A: To be implemented in 1997). Upgrading of regional rail system to suburban mass transport (I: infrastructure development in effect since 1995).	Subsidies for building of urban monorails/new transport systems (I: 11 lines built in fiscal 1992). Fiscal incentives to encourage use of shipping and rail over road transport for freight: including reduction of fixed asset taxes on private containers for railway transportation; loans from Japan Development Bank to assist in providing facilities for cargo handling, storage, loading and unloading, container storage and for container chassis pools to store trailers (I).

Table 4c. **Economic Instruments** cont'd

Policies/Objectives cited	Hungary	Ireland	Italy	Japan
Use of subsidies for traffic management systems		Fee assessment on rush-hour driving (A: to be introduced 1996-97). Better traffic management and enforcement in Dublin. Increased number of bus lanes; extension of operating time of existing buses; expansion of closed circuit television monitoring; promotion of car pooling (I).	Increase in use of teleconferencing to reduce need for travel (I: infrastructure development under way since 1993). Installation of bookable traffic lights at remote crossings (A: infrastructure development to be implemented in 1998).	• Establishment of new bus lanes, automatic signalling, prioritising systems for prompt service, control of illegal parking. Subsidies for traffic management, bus location and general information systems. (I: 16 projects implemented in fiscal year 1992). • Five-Year Project for Traffic Management Systems Installation. (I: Yen 2,015 trillion from 1991-1995. Traffic control centres and signals upgraded in 1993; illegal parking control systems, travel time measurement information systems).
Use of subsidies to encourage use of alternative-powered vehicles / new technologies	Approx. US$2.7 million invested by Central Environmental Fund in 1994 for construction of LPG/CNG stations (I: number of LPG-powered cars has increased over last years).	Introduction of new, energy-efficient buses in Dublin fleet. (I: increase in share of commuter travel into the city centre by bus from 22% in 1990 to 24% in 1991).	Introduction of high-efficiency engines and continuous variable transmission; application of improved technology (A: scheduled to be implemented in 2000.) Introduction of start-stop system for engines and reduced pumping (A: scheduled to be implemented in 1996.) Application of new technologies and stock renewal for buses and freight lorries (I: In effect since 1995 for buses and since 1996 for freight lorries).	• To increase use of low-emission vehicles in private company fleets: subsidy from the Pollution-Related Health Damage Compensation & Prevention Association for the purchase of electric vehicles, lease of methanol-fuelled vehicles, and installation of methanol filling stations. (I: fiscal 1992: subsidies granted for 63 electric vehicles, 156 methanol-fuelled vehicles (including extension of leases), and one methanol filling station.) • Subsidy to families for acquisition/lease of CNG-fuelled vehicles, hybrid engine-vehicles and fuelling equipment. (I: Since 1994.) • Eco-Station 2000 Plan: Subsidies for petrol service stations that install fuel supply equipment for electric, CNG- and methanol-fuelled vehicles along with conventional fuel pumps. (I: in effect since fiscal 1993).

Table 4c. **Economic Instruments** cont'd

Policies/Objectives cited	Hungary	Ireland	Italy	Japan
Fiscal measures to promote sale of more fuel-efficient vehicles		Vehicle Registration Tax on purchase of new cars (I: reduced in 1994 as measure to boost automotive industry: tax of 23.2% for cars of 2 500 cc; 29.25% for cars over 2 500 cc). Graduated road tax regime based on engine capacity (larger cars subject to higher tax levels). (I: tax ranges from IR£ 92 for cars up to 1 000 cc to IR£ 800 per year for cars greater than 3 000 cc).	Increase in property tax of cars scaled to engine size (status not communicated). Status in 1996: luxury tax of L 5 to 10 million according to engine size; transfer taxes at flat rate in addition to VAT; flat rate registration charge.	• National subsidy and local tax provisions for introduction of low-emission vehicles as environmental pollution patrol cars; municipal bond funding and local tax provisions for introduction of low-emission waste-collection vehicles and municipal buses. (I: 50 low-emission vehicles put in operation for postal service between fiscal 1992 and 1993; 62 low-emission environmental pollution patrol cars introduced in fiscal 1992; 25 low-emission municipal buses introduced in fiscal 1992). • Reduction of annual automobile ownership taxes and acquisition tax for all owners of CNG-fuelled vehicles, hybrid engine and methanol-fuelled vehicles (I). • Fiscal incentives for energy investment available for lightweight rolling stock cars, ship boilers that use waste heat, etc. (I).
Vehicle fleet renewal	Forint 30 000 in public transport tickets provided in exchange for a Trabant vehicle as an incentive to retire old two-stroke vehicles from fleet (I: 10 000 vehicles traded in under programme in 1993; 100 000 retrofitted with subsidised catalysts).		Initiatives to increase efficiency of car fleet (I: In effect since 1993). Incentives for better maintenance of cars less than 5 years old; more than 5 years old (A: To be implemented in 1996).	

Table 4c. **Economic Instruments** cont'd

Policies/Objectives cited	Hungary	Ireland	Italy	Japan
Fiscal measures to encourage use of public transport for professional commuter traffic / reduce incentive for business travel		Increase in benefit-in-kind taxation of company cars to provide a fiscal disincentive for company car use. (I: Increase in car value threshold for calculating capital allowances and expenses increased to Ir£ 10 000 in 1992; this ceiling deemed too restrictive/fiscally putative in 1994 -- it was therefore increased to Ir£ 13 000 and then Ir£ 14 000 in 1995; in 1996, 20% relief from benefit-in-kind taxation introduced for company representatives spending 70% or more of time travelling on business, where business mileage exceeds 5 000 miles per year. Concession to cost Ir£ 1.7 million per year).		
Other	EC-Phare Programme funding overhaul of 150 diesel bus engines in Budapest, Györ and Pécs (I); Central Environmental Protection Fund extension of same programme; EBRD financing for rebuilding of 500 bus engines (I); Budapest municipality planning to retrofit 300 buses for CNG (A).		Thermie Programme -- Case Study of Bologna: integrated mobility management including smart traffic lights; restricted lane control; bookable bus stops; integrated ticketing, variable signalling (I). Thermie Programme -- Case Study of Florence: Redesigning of public transport network; new rail lines; additional ring roads and interchange areas; restricted access zones to encourage modal shift (I).	

Table 4d. **ECONOMIC INSTRUMENTS**
Latvia, Lithuania, Netherlands, New Zealand

Policies/Objectives cited	Latvia	Lithuania	Netherlands	New Zealand
Fuel taxes to discourage excessive use of road vehicles		Optimisation of fuel prices/establishment of fines for pollution (I : initial research complete; implementation under way in association with the Harvard Institute for International Development).	• Road fuel tax increase of 15 Dutch cents ($0.9) per litre (7.5%) for petrol and 5 cents ($0.3) per litre (3.5%) for diesel: announced by government in September 1996, to be offset by decrease in car ownership taxes (measure consistent with Dutch policy of discouraging car use rather than car ownership.) (A: to be implemented mid-1997). (*Environment Watch*, September 1996). • Increase in excise tax (1993-2000) (I, A: 20 cents/litre between 1993-2000; 11 cents increase already in place (excluding inflation), in addition to 7% autonomous price increase for petrol; 14% for LPG, diesel). • Increase in excise taxes (1986-1993) to raise the real costs of private car use (I: 35% increase in excise taxes between 1986 and 1993 – inflation-adjusted).	Central government has recently discontinued a regional tax introduced in 1992 and returned to central system in order to maintain public transport services (I).
Fuel tax exemption/ reduction to promote use of certain fuels	Differentiated excise tax to encourage use of less-greenhouse gas emitting fuels (P).			
Government subsidies / other support to encourage use of public transport / inter-modal transport / modal shifting / discourage excessive road transport	Improvements in water and railway transport networks, services and speed capacity to promote use of these modes (P). Improvements to public transport system (P).	Development of trolley-bus network in Vilnius and Kaunas ; organisation of network in Klaipeda (I : extension of lines / improvements to rolling stock on annual basis).	• Development of regional public transport networks / light rail (I). • Introduction of high-speed rail lines from Amsterdam to Belgium and Germany (I). • Rail 21: Dutch Railways planning scheme up to 2010. Improvement in connections, capacity, service and speed. Total rail investment 1988-2010 more than F 20 billion (I).	

Table 4d. **Economic Instruments** cont'd

Policies/Objectives cited	Latvia	Lithuania	Netherlands	New Zealand
Use of subsidies for traffic management systems		Support for road and traffic safety development via the Road and Traffic Safety funds (status not clear).	Subsidies to local governments and companies for Traffic Demand Management (employer trip reduction) and local infrastructure measures and facilities e.g. cycling paths (I).	
Use of subsidies to encourage use of alternative-powered vehicles / new technologies			SSZ Technology Programme: subsidies for Research and Development of alternative fuelled electric and hybrid vehicles (I).	
Road taxes and road pricing			Road pricing to be introduced after 2000 (P).	
Fiscal measures to promote sale of more fuel-efficient vehicles			Shift in emphasis on vehicle-related taxes instead of road taxes to stimulate purchase of cleaner, more efficient cars (I: introduced in 1995). Increase in annual tax on LPG vehicles (A).	
Fiscal measures to encourage use of public transport for professional commuter traffic / reduce incentive for business travel			Fiscal benefits for Traffic Demand Management participants / companies (I).	
Tariff adjustment to encourage public transport use.			Increases in public transport tariffs, not to exceed variable car costs (I).	
Other		Replacement of steam with electricity for locomotives (I: In 1996, two steam generators replaced).		

Table 4e. ECONOMIC INSTRUMENTS
Norway, Poland, Portugal, Romania

Policies/Objectives cited	Norway	Poland	Portugal	Romania
Carbon or CO_2 tax	CO_2 tax (I: since 1991. Tax rate in 1996 = 0.83 NKr/litre petrol and 0.415 NKr/litre diesel. Petrol consumption dropped by more than 5% 1990-1993; however, economic downturn may have contributed to this).			
Fuel taxes to discourage excessive use of road vehicles		Fuel taxes, feebates to promote use of less polluting vehicles and fuels (E).		
Government subsidies/ other support to encourage use of public transport / inter-modal transport / modal shifting / discourage excessive road transport	Support to Norway's four largest cities via subsidies to public transport investments (I).	Promote energy efficient transport means [modernisation of railways (also R/G)] (A). Investment in urban public transport (P).	Reinforce competitive position of rail transport relative to private transport; implementation of rail activity plan (I: Public investment in rail infrastructure in progress as of 1996). Development of infrastructure for modal transfers: improve capacity and service of suburban railways; implement network expansion plan for Lisbon underground; launch light rail service in urban areas other than Lisbon; introduce high-capacity articulated trams (I: work in progress through 2005). Inter-modal transport organisation: construction/ re-organisation of stations and passenger interfaces and platforms for combined transportation network (I: in progress 1995-1998).	Economic incentives and research and development envisaged to encourage shift to public transport use (E, also R&D).
Use of subsidies for traffic management systems			Modernisation of road infrastructure; improvements to safety and traffic flow. Completion of National Road Plan infrastructure development; building of bridge over River Tagus in Lisbon (I: work underway in stages through 2000).	

79

Table 4e. **Economic Instruments** cont'd

Policies/Objectives cited	Norway	Poland	Portugal	Romania
Use of subsidies to encourage use of alternative-powered vehicles / new technologies	Subsidies for research and development of alternative fuel sources; particularly, natural gas use in buses and ferries (I).	Government action to encourage introduction of "modern" pollution reduction technologies (P) (also R/G, EI, R&D).		
Road taxes and road pricing	Use of road pricing as a means of increasing the price of road transport and thereby reducing private transport demand (E: under consideration).	"Infrastructure" and "ecological" fees to contribute to internalisation of external costs (P).		
Fiscal measures to promote sale of more fuel-efficient vehicles	Differentiated purchase tax on private cars based on engine power, engine volume and weight of car (I: in effect).			

Table 4f. **ECONOMIC INSTRUMENTS**
Slovak Republic, Spain, Sweden, Switzerland

Policies/Objectives cited	Slovak Republic	Spain	Sweden	Switzerland
Carbon or CO$_2$ tax			Tax on emissions of CO$_2$ in addition to general energy and environmental taxes and broadening of VAT to include transport (I: As of 1996, SKr 0.86 for petrol, SKr 1.05 for diesel).	
Fuel taxes to discourage excessive use of road vehicles	Differentiated fuel tax favouring diesel. As of May 1995, Sk 9390 to 10 800/t for petrol and Sk 8 250/t for diesel. LPG : Sk 2 370/t; CNG : Sk 2/m^3 (I).		General energy and environmental taxes: SKr 3.3 for petrol, approx. SKr 2.0 for diesel. Gradual increase in taxation from 1990-1996 (I).	Increase in fuel prices via a one-off increase in import duty to bring the price of fuel in Switzerland in line with that of neighbouring countries (I: 20 centimes per litre increase 8 March 1993, leading to a projected 5 to 10% drop in fuel consumption from road transport).
Fuel tax exemption / reduction to promote use of certain fuels	Slovak National Council Law No 87/1994:Road tax exemption on electric- or solar-powered vehicles for five-year period; road tax reduced by 50% for LPG/CNG-powered vehicles (I: 3 public filling stations for LPG now available. In 1995, approx. 300 passenger cars and 20 urban mass transit bus engines retrofitted for CNG).			
Government subsidies / other support to encourage use of public transport / inter-modal transport / modal shifting / discourage excessive road transport	• Subsidies (investment and non-investment) of mass and railway transit (I: in 1993: Sk 1.73 billion to railway transport; Sk 2.41 billion to public road transport. In 1994, Sk 3.17 billion to railway transport, and 1.96 billion to public road transport.) • Decree No. 499/95: subsidies from state budget for transport infrastructure; in particular, electrification of railways and urban transport [I: 39.1% of railway is now electrified (1 430km). Electrification of further 120 km of rail up to 2000 – approx. Sk 1.2 billion investment. All principal rail routes to be electrified by 2005. Electric traction now in transport systems of 5 cities]. • Improvements to infrastructure to promote cycling (I/E : more cycling paths are being built, but progress is insufficient).	Improvements to urban transport infrastructure (status not communicated).		Federal support for combined transport and accompanied motor vehicle transport in the form of cost recovery compensation in combined transport, price subsidies for accompanied motor vehicle transport, and investment financing (I).

Table 4f. **Economic Instruments** cont'd

Policies/Objectives cited	Slovak Republic	Spain	Sweden	Switzerland
Use of subsidies for traffic management systems	• Traffic control via co-ordinated systems of traffic lights at main cross-roads (status not communicated) • Progressive parking charges in urban centres and main tourist sites (I: in effect in main Slovak cities and tourist regions).		SKr 500 million spent from 1993-1996 on programmes (excluding research) to promote a more "energy-efficient, climate-friendly, transport structure" (I).	
Use of subsidies to encourage use of alternative-powered vehicles / new technologies		Introduction of new urban buses fuelled by LPG and natural gas (I).		
Road taxes and road pricing	On-road motor vehicles subject to taxation based on engine volume (I: tax ranges from 1 Sk 1 200 per 900 cm^3 to Sk 3 600 per cm^3 if over 3 000 cm^3. Tax rate on utility cars and buses based on total weight ranges from : 1t - Sk 1 200 to Sk 54 000 for over 40t. Tax is irrespective of vehicle age). Exemptions include: public transit vehicles, urban waste removal vehicles, vehicles powered by solar or electric energy.			
Fiscal measures to promote sale of more fuel-efficient vehicles	Measure to be integrated in state budget preparation for 1998 (A).			

Table 4f. **Economic Instruments** cont'd

Policies/Objectives cited	Slovak Republic	Spain	Sweden	Switzerland
Vehicle fleet renewal	Waiver of customs and import charges on new passenger cars with an engine volume less than 1 500 cm³ (I).	"Plan Renove": Ptas 100 000 paid for vehicles over 10 years old scrapped and traded for new vehicles (I: number of newly registered vehicles increased over 20% in the first 8 months of 1994 relative to same period in previous year). "Plan Renove 2": From 10-94 to 06-95: Ptas 80 000 paid for vehicles over 7 years old scrapped and traded for new vehicle (I).		
Tariff adjustment to encourage public transport use		Encourage use of monthly transit passes (status not communicated).		
Actions to support combined transport				Federal compensation to Swiss rail companies for: cost recovery in rail transport of trailers; investment subsidies to rail companies and other parties; reduction in price of accompanied motor vehicle transport (I).

Table 4g. **ECONOMIC INSTRUMENTS**
United Kingdom, United States, European Union

Policies/Objectives cited	United Kingdom	United States	European Union
Carbon or CO_2 tax			Proposal for carbon tax to be introduced in steps and modulated 50% according to energy content and 50% according to carbon content of fuels (P: proposed in 1992; negotiations on voluntary CO_2/energy taxes stumbled on opposition from Member states in late 1995).
Fuel taxes to discourage excessive use of road vehicles	Fuel duty strategy introduced in 1993 in response to targets agreed at Rio in 1992. Government announced intention to raise road fuel duties by an average of at least 3% a year in real terms; this was later strengthened to 5% a year in real terms (I).		(See Regulations and Guidelines: Strategy element on fiscal measures).
Fuel tax exemption / reduction to promote use of certain fuels	The UK maintains a differential in fuel duty to favour unleaded petrol, and duty on road fuel gas was recently cut for the second year running to encourage the use of gas-powered vehicles (I). Ultra-low sulphur diesel is also to be taxed at a lower level relative to ordinary diesel (P).		
Government subsidies / other support to encourage use of public transport / inter-modal transport / modal shifting / discourage excessive road transport	Targeted support for rail freight via the long-established freight facilities grant, increased in 1993 by the track access grant. Since 1979, some 146 grants have been awarded, for a total value of around £100 m at today's prices (I).		

84

Table 4g. **Economic Instruments** cont'd

Policies/Objectives cited	United Kingdom	United States	European Union
Use of subsidies to encourage use of alternative-powered vehicles / new technologies		Federal income tax deductions for alternative-fuelled vehicles: e.g. up to US$200 for purchase/conversion of "clean-fuel" cars; US$5 000 for light trucks, US$50 000 for heavy-duty trucks and buses, US$100 000 for refuelling stations/equipment; up to US$4 000 in tax credits for electric vehicles. Many states offer additional vehicle tax deductions, rebates or credits. Also substantial federal and state tax credits for certain alternative and /or renewable motor vehicle fuels (I: since 1992).	
Fiscal measures to promote sale of more fuel-efficient vehicles		"Gas-guzzler" tax, US$1 000- 7 700 increasing with fuel consumption for cars below 22.5mpg (I) (Not part of the Climate Change Action Plan).	
Fiscal advantages for public transport companies	Fuel duty rebate for bus operators to help keep services running (I).		
Fiscal measures to encourage use of public transport for professional commuter traffic / reduce incentive for business travel		Reform of federal tax subsidy (US$65/month/parking space) for employer-provided parking to « level the playing field » among travel modes ; encourage commuters to use public transport, car-pooling or other means for travel to work (P : mechanism known as «parking cash out», allowing employees to choose between US$65 cash in taxable income or parking place. In 1993, a proposal for a mandatory parking cash-out system failed to make it to Congress for vote; a voluntary cash-out system now being proposed).	

Table 5a. **REGULATIONS AND GUIDELINES**
Austria, Belgium, Czech Republic, Denmark

Policies/Objectives cited	Austria	Belgium	Czech Republic	Denmark
Heightened speed limit enforcement to decrease fuel consumption; increase safety	Improved speed limit enforcement to decrease fuel consumption by 1 to 1.5 l/100 km, and increase safety (I - 1 October 1994). Mandatory use of electronic speed limit devices for HGVs of 12 tonnes or more (85km/h); omnibuses of more than 10 tonnes (100km/h); 60km/h speed limit for low-noise HGVs at night (I: 1 January 1995).	Government plan to secure commitment to speed limit enforcement from all actors (road user groups, car manufacturers, insurance companies); co-ordination between different public authorities; driver and youth education; driver sensitivity and responsibility campaigns; enforcement of speed infraction rules (P).		
Traffic management; restrictive policies to discourage private vehicles in city centres / reduce fuel consumption.	Night-driving ban for "non-low noise-HGVs (see above) (I).	Parking restrictions / digressive parking tariffs: high in city centre decreasing in periphery (P).	Optimisation of road traffic in selected areas (I).	
Vehicle fuel consumption standards			Implementation of specific fuel consumption targets for vehicles in development (I) (A).	
CO_2 emissions standards			Setting of CO_2 emissions standards for air transport (I) (A).	

Table 5a. **Regulations and Guidelines** cont'd

Policies/Objectives cited	Austria	Belgium	Czech Republic	Denmark
Development of programmes/ guidelines on climate change / transport and environment	Five Points Programme of the Austrian Government: includes measures to raise the competitive position of public transport; promote new technology (I: in progress). 1991 Master Transportation Concept (I: continued implementation. Includes regulations, regional / land-use planning, infrastructure investments).		Programme of stabilisation and reduction of CO_2 emissions from transport in the Czech Republic (I: 1993). Measures designed to: - decrease fuel consumption - rationalise use of infrastructure and transport - promote energy-saving modes of transport - extend combined freight transport. (specific measures included in these tables)	• Action Plan on Transport 1990 (I: Long-term plan for environment and development aimed at achieving sustainable transport policy). • Transport 2005 (I: 1993; reviewed implementation of energy and environmental targets of Action Plan). • Action Plan for Reduction of the CO_2 Emissions of the Transport Sector (I: May 1996; follow-up to Transport 2005, identifies means and instruments for reducing CO_2 from transport. Government states intention to support development towards greater energy efficiency and a reduction in the need for transport, as well as emphasising bicycle and public transport, etc.).
Mandatory transport plans for companies/ measures to promote telecommuting		Proposed law to render obligatory transport plans for companies with over 50 employees (P).		
Other	Ordinances and recommendations to promote use of biofuels in ecologically sensitive areas (A: partly implemented; partly planned).		Conclusion of international agreements aimed at regulation of road freight transport (I).	

Table 5b. **REGULATIONS AND GUIDELINES**
Finland, France, Germany, Hungary

Policies/Objectives cited	Finland	France	Germany	Hungary
Heightened speed limit enforcement to decrease fuel consumption; increase safety		Introduction of tamper-proof tachograph; regulation of drivers' working hours, total weights and speeds (A: Adopted by Parliament in January 1995; presented at EU level).		
Mandatory vehicle inspections	Mandatory annual vehicle inspections. For new passenger cars: on their third and fifth year; thereafter on an annual basis. Passing the inspection also requires passing the emission test (for diesel cars as well). Emission limits and measurements follow Directive 92/55/EEC (I).	Mandatory vehicle revision following inspection for road freight vehicles to be extended to ensure compliance with pollution emission standards (A: Planned for 1 January 1997). For passenger transport, strict enforcement of regulations requiring mandatory repairs and maintenance to ensure compliance with emission standards following inspection (I: As of 1 January 1995, technical inspection is required every two years for vehicles over 4 years old).		Carried out at licensed service stations; owners whose vehicles do not meet standards have 15 days to make improvements (I: since 1992: number of vehicles passing yearly emissions tests has increased; average emissions' level has dropped).
Development of programmes/ guidelines on climate change / transport and environment	1994 action programme for transport: proposes long-range goals, targets and measures for 2000 (E).			National Energy Efficiency Improvement and Energy Conservation Programme 1991: re: Transport (see vehicle taxation) (I: As of 1993, 2 regulations in effect on annual taxation based on weight of cars; up to 1 000 kg, taxation relatively low; above that weight, taxation higher).
Measures to promote public transport	Working group established to develop a public transport programme. (I) This, with a view to improving the price and service competitiveness of public transport (increased competition, new information technology) (E).			

88

Table 5b. **Regulations and Guidelines** cont'd

Policies/Objectives cited	Finland	France	Germany	Hungary
Efforts to integrate transport and land-use planning	Sectoral planning re-organisation to account for transport in land-use planning (status unclear).			
Import restrictions on highly polluting vehicles (especially in CEE countries)				Import of vehicles more than six years old and of vehicles with two-stroke engines prohibited (I: since 1995; drop in number of old, imported cars. As of 1996, age limit is no more than four years old). Emissions standards for new and imported used petrol vehicles equivalent to those introduced in EU in 1993 (I: since 1995). Application of strict emissions standards for new diesel vehicles, including buses and trucks (UN-ECE standard 49.2) (I: since 1995).
Other			Railway structural reform to create a framework for the improvement of railway performance and competitivity: render rail transport more flexible; encourage modal shift to rail. Involves privatisation of Deutsche Bundesbahn (German Federal Railway) and Deutsche Reichsbahn (rail system in ex-GDR) (I: In effect since 1 January 1994). Regulation requiring vapour return systems at petrol stations (I: since 1993, all filling stations have to be retrofitted with equipment reducing emissions when fuelling a vehicle).	

Table 5c. **REGULATIONS AND GUIDELINES**
Italy, Japan, Latvia, Lithuania

Policies/Objectives cited	Italy	Japan	Latvia	Lithuania
Heightened speed limit enforcement to decrease fuel consumption; increase safety	Speed limits and efficient guide stiles outside of cities (I: in effect since 1995).		Strict enforcement of speed limits (I).	
Mandatory vehicle inspections			Regulations in place requiring annual vehicle inspections (I).	
Traffic management; restrictive policies to discourage private vehicles in city centres/ reduce fuel consumption.			Legislation restricting private transportation in cities (I: currently being implemented).	
Vehicle fuel consumption standards		• Fuel-efficiency targets for 2000 set: average 8.5% improvement over fiscal 1992 levels (I: 1979 standards tightened in 1993). • 5% target for average improvement in fuel efficiency for petrol trucks (I).		
Development of programmes/ guidelines on climate change/ transport and environment			National Programme for Transport Development (1995-2000) (I); State Programme for Development of Transport (1994) (I: details not provided).	Comprehensive transport and environment strategy being prepared in 1996 with support from EU-Phare Programme, Sweden and the Vilnius urban development programme (I: development in progress).
Measures to promote combined transport		Law Concerning Construction of Distribution Business Centres; provides for low-interest loans and guarantees from Japan Development Bank for establishment, co-ordination and operation of joint delivery services; tax credits and low-interest loans from JDB and others for multi-functional distribution centres (I: Two construction projects underway as of fiscal 1992. System reviewed and extended to additional cities in 1993.		

Table 5c. **Regulations and Guidelines** cont'd

Policies/Objectives cited	Italy	Japan	Latvia	Lithuania
Other				Preparation of national environmental standards for the transport sector/harmonisation with EU standards (I: two standard documents produced in co-ordination with Ministry of Environment in 1996 for petrol and diesel engines based on EU directives; standard on vehicle noise being considered). 1996 law on Environmental Impact Assessment (I).

Table 5d. **REGULATIONS AND GUIDELINES**
Netherlands, New Zealand, Norway, Poland

Policies/Objectives cited	Netherlands	New Zealand	Norway	Poland
Heightened speed limit enforcement to decrease fuel consumption; increase safety	Comprehensive speed limit enforcement and communications programme, including upgrading of police output and enforcement tactics. Goal: Reduce average motorway speed to 1983 level of 106 km/h (1987: 112; 1995: 110). (I: since 1988. In 1995, over 600 000 fines were given for motorway speeding as compared to 60 000 in 1987. Effect of this policy was a 360 000 tonne reduction in CO_2 or 1.5% of road traffic emissions).	Improved speed limit enforcement and speed limit education (I: major reductions seen in open road speeds in the first half of 1994 following introduction of speed cameras).	Minimisation of maximum speed limits for cars (I: speed limits in Norway are relatively low compared with other European countries).	
Mandatory vehicle inspections	Mandatory verification of engine adjustments as part of annual vehicle inspections (I: legislation promulgated in 1991).		Application of EU vehicle inspection system in order to stimulate improvements in average vehicle energy efficiency (I: in effect).	
Traffic management ; restrictive policies to discourage private vehicles in city centres/ reduce fuel consumption	Transport demand management measures including company TDM plans for employees (car-pooling, collective contracts with public transport companies; improved provisions for cyclists, awareness campaigns) (I).			

Table 5d. **Regulations and Guidelines** cont'd

Policies/Objectives cited	Netherlands	New Zealand	Norway	Poland
Development of programmes/ guidelines on climate change/ transport and environment	Integrated Transport and Environment Policy Programme (SVV/NMP) linking environmental policy goals for climate change to transport policy planning and measures. Programme based on road traffic emission reduction targets for No$_x$ (75%) and CO$_2$ (10%) 1986-2010. Policy specifies setting short-term priorities for better speed limit enforcement (I).	Regional Transport Strategies (I: Research underway in 1992 on how best to incorporate CO$_2$ into these strategies. Status not communicated). National Vehicle Fleet Strategy: designed to manage impact and structure of New Zealand's vehicle fleet; reduce emissions; improve fleet fuel efficiency (Draft strategy was to be presented to Government in mid-1995 before release to public for consultation. Status not communicated).	National policy guidelines elaborated under the Planning and Building Act of 1985 (I: Adopted in 1993).	Establishment of priorities for public transport (E) (also VA, IN).
Efforts to integrate transport and land-use planning	Strict application of government land-use policy to encourage new companies and residential projects to locate in appropriate areas (I). "ABC" location/parking capacity ruling for built-up areas (I).	Preparation of integrated land transport strategy, accounting for regional transport needs, safety, cost and environmental considerations (status not communicated).		Design of national, regional and local planning schemes to avoid urban sprawl and excessive growth in transport demand (E) (also VA).

Table 5e. REGULATIONS AND GUIDELINES
Portugal, Russian Federation, Slovak Republic, Sweden

Policies/Objectives cited	Portugal	Russian Federation	Slovak Republic	Sweden
Heightened speed limit enforcement to decrease fuel consumption; increase safety	Highway Code modified by Decreto-Lei n° 114/94: speed limits lowered for certain classes of vehicles to encourage more responsible driving behaviour; penalties significantly tightened for infractions (I).		Enforcement of 80 km/h speed limit in certain areas (E: implementation envisaged only in case of excessive future CO_2 emissions).	
Mandatory vehicle inspections	Periodic vehicle inspection; introduction of new technologies via renewal and modernisation of in-circulation vehicle fleet (I: In effect for all classes of vehicles since 1993; inspection systems modified in April 1996 (Portaria n° 117A/96).		Federal Ministry of Transport regulations mandating inspection of vehicles for adherence to exhaust emission standards and technical soundness. (I : network of 48 technical inspection and 450 emissions control stations in place).	
Traffic management ; restrictive policies to discourage private vehicles in city centres / reduce fuel consumption	Use of automation and designated lanes for public transport for improved traffic management (I: since 1994, use of automatic highway toll system; "Via Verde", which allows drivers to pass through toll stations without stopping and pay via direct banking).		Measures to improve effectiveness of urban traffic management (parking systems, creation of pedestrian zones). (I : in limited application in Bratislava and larger cities).	
Vehicle fuel consumption standards		Development of vehicle fuel efficiency standards (E).		Target for private car average fuel consumption of 0.63 litres per 10 km as of 2005 has been proposed in the context of the Committee on Transport Policy (P). (The Volvo Co. has committed to a 25% reduction in average fuel consumption by 2005).
CO_2 emissions standards		Development of standards on CO_2 emissions from vehicles (E).		

93

Table 5e. **Regulations and Guidelines** cont'd

Policies/Objectives cited	Portugal	Russian Federation	Slovak Republic	Sweden
Development of programmes/guidelines on climate change / transport and environment	Global restructuring of rail transport sector; creation of three working groups, the objective of which is to carry out technical and economic studies; develop legislation and an action plan for a new regulatory body; create a new rail transport company and a rail infrastructure management company (I: work of three groups under way).	Creation of inter-ministerial commission on climate change (I); development of proposals to meet the Russian Federation's commitments to FCCC (P). Development of State Programme on greenhouse gas emissions reduction (A) focusing on transport (P).		Creation of government-appointed Transport and Climate Committee to examine use of alternative fuels, public transport and the setting up of "environmental zones" in larger cities (I: final report produced in May 1995). Appointment of Government Committee on Transport Policy to draw up national plan for transportation in Sweden (I: final report to be produced at end of 1996).
Measures to promote combined transport			Strategy to develop combined transport with a view to limiting road traffic (I: system based on ISO IC large containers. 1 380 km of railway to Slovak Republic. Transhipment yards in Bratislava, Zilina, Kosice and Cierna nad Tisous. Further development envisaged).	
Efforts to integrate transport and land-use planning	Integration of transport in multi-sectoral plans for territorial development; regulatory improvements in this area (I: merging of ministries of Transport and Territorial Development by Decreto-Lei n°23/96).			

94

Table 5f. REGULATIONS AND GUIDELINES
Switzerland, United Kingdom, United States, European Union

Policies/ Objectives cited	Switzerland	United Kingdom	United States	European Union
Heightened speed limit enforcement to decrease fuel consumption; increase safety	Ordinance setting technical requirements for road vehicles: speed limitation devices are obligatory for HGVs in use for the first time as of 1/1/96. For HGVs in use for the first time between 1/1/88 and 31/12/95, the devices are required as of 1/1/98. Regulated speeds are based on EU Council Directive 92/6 of 10/2/92 relative to the installation and use in the EU of speed limitation devices on certain categories of motor vehicles (I).			Council Directive 92/6 EEC of 10.2.1992 on speed limitation devices for HGVs and buses; provides for installation of speed limitation devices on new HGVs and buses from January 1994, and retroactively on HGVs and buses first registered from 1988 to 1993 as of January 1995. Maximum speeds set at 100 km/h for buses, 85 km/h for HGVs. Vehicles used only for national transport affected by this at the latest in January 1995 (I).
Mandatory vehicle inspections	Road Traffic Ordinance stipulating that the following are subject to inspection/ maintenance with regard to tailpipe exhaust (including CO, emissions in particular): 1) light vehicles equipped with electronic-ignition engines and speed capacity of minimum 50 km/hr: with catalyst, every 24 months; without catalyst, every 12 months 2) vehicles equipped with compression-ignition engines with speed capacity of more than 30 km/hr: every 24 months 3) vehicles equipped with compression-ignition engines with speed capacity of max. 30km: every 48 months (I).		Periodic inspection and maintenance (I/M) of cars and light trucks required in areas that do not meet federal air quality standards. Vehicles must meet "basic" or "enhanced" performance standards, depending on the severity of the local air quality problem. Each state develops its own programme to meet standards (I).	
Traffic management; restrictive policies to discourage private vehicles in city centres / reduce fuel consumption		Planning guidance to local authorities to reduce the need to travel, action on speed limits and their enforcement (I).		

Table 5f. **Regulations and Guidelines** cont'd

Policies/Objectives cited	Switzerland	United Kingdom	United States	European Union
Vehicle fuel consumption standards	Federal government Ordinance on Reducing the Specific Fuel Consumption of Cars (I: since 1 January 1996, fuel economy target for newly registered vehicles of a 15% reduction in average fuel consumption 1996-2001: 3.2% per year).		Corporate Average Fuel Efficiency (CAFE) standards (I: Established in 1975; went into effect for automobiles in model year 1978 and for light trucks in model year 1979. Revised most recently in 1992. Currently: 27.5 mpg. Pre-dated the Climate Change Action Plan).	
Development of programmes/ guidelines on climate change / transport and environment	As part of the national Energy 2000 Programme : a number of government-supported projects and programmes designed to increase energy efficiency in passenger and freight transport. Annual budget: SF 6 million with private sector co-financing. Two measures of particular note for transport and CO_2: new taxation of HGVs and implementation of the "Alpine Initiative" (P).	The UK Climate Change Programme (1994) set out the range of measures to limit greenhouse gas emissions across all sectors of the economy (I).	Adopt a Transportation System Efficiency Strategy to promote state adoption of measures that limit growth in vehicle travel. Includes: R/G: transportation conformity rule to ensure that state clean air implementation plans are consistent with state transportation infrastructure plans; EI: credits for emissions reductions under Clean Air Act; ISTEA funding; Federal outreach. States will use measures including: market mechanisms to encourage less driving; parking cash out; transit subsidies (I).	Strategy for reducing CO_2 emissions from passenger cars; Commission communication COM (95)689 final, 20.12.1995; Council conclusions of 25.6.1996. Strategy including VA; EI, IN: proposals on monitoring and fuel-economy labelling (to be presented by Commission in 1997) and fiscal measures (under study in 1996). Commission invited by Council to report on other measures for CO_2 abatement in transport in 1997. Objective is to achieve an average of 120g/km CO_2 emissions (approx. equiv. to 5 l/100 km fuel consumption for petrol and 4.5 l/100 km for diesel cars) for new cars by 2005. (A: political agreement between Council and Commission on overall objective and approach; strategy to be implemented 1996-1997).
Mandatory transport plans for companies / measures to promote telecommuting		Government consulting with Transport 2000 and local authorities among others about the potential for Green Commuter Plans -- a national good practice guide could be developed building on local experiences (I: ongoing).		

Table 5f. **Regulations and Guidelines** cont'd

Policies/Objectives cited	Switzerland	United Kingdom	United States	European Union
Efforts to integrate transport and land-use planning		Planning Policy Guidance (PPG) 13 on Transport: aims to achieve better integration of transport and land-use planning so as to reduce growth in length and number of motorised trips, encourage alternative means of travel and reduce reliance on the private car (I: Guide to better practice published in October 1995; implementation being monitored).	(See transportation conformity rule in box above).	
Other	Referendum of 27 September 1992 in favour of the construction of a trans-alpine rail link contributing to a European high speed network and encouraging a shift in HGV traffic to rail (first links set to open in 2005, 2007) (I).		"Car Talk": multi-stakeholder advisory committee established in 1995 to explore ways of reducing GHG from transport. (I: Consensus not reached within group, but significant analysis produced; part of Climate Change Action Plan).	

Table 6a. **VOLUNTARY AGREEMENTS/ACTIONS**
Austria, Canada, Czech Republic, France

Policies/Objectives cited	Austria	Canada	Czech Republic	France
Agreements with vehicle manufacturers	Agreement with car manufacturers to improve vehicle fuel efficiency to 3-litre/100 km (E).	Motor Vehicle Fuel Efficiency Initiative: Voluntary agreement under separate Memoranda of Understanding with each of the Canadian motor vehicle manufacturers (1995) and the US Department of Energy (1996) on increasing fuel efficiency of new vehicles; possibility of higher new vehicle fuel efficiency standards (A).	Increase in production and use of alternative fuels (A).	Government-industry working group established to explore the possibility of a Europe-wide limit on truck engine power with a view to reducing energy intensity by 20% in 2015 (I). French car manufacturers have set a target of cutting average CO_2 emissions from their vehicles to 150g per km by 2005 (A).
Alternative fuels			Use of alternative fuels in urban public transport vehicles (I) (A).	
Combined transport			Implementation of combined transport actions in selected areas (I).	

Table 6b. **VOLUNTARY AGREEMENTS/ACTIONS**
Germany, Lithuania, Netherlands, Sweden

Policies/Objectives cited	Germany	Lithuania	Netherlands	Sweden
Agreements with vehicle manufacturers	Agreement with domestic vehicle manufacturers on fuel economy: calls for a 25% reduction in average fuel consumption of cars between 1990 and 2005 -- a rate of 1.9% per year (I).			Swedish car manufacturer VOLVO to lower average fuel consumption of its cars sold in the EU by 25% from 1990 to 2005 (I).
Other		Agreement between Klaipeda port authority and electricity network for supply of electricity to ships by city (E: plan to be designed after privatisation completed).	Agreement with the freight transport and trucking business on reducing CO_2 and other emissions through logistics, efficiency and technology measures (I: since 1993).	

98

Table 6c. VOLUNTARY AGREEMENTS/ACTIONS
United Kingdom, United States, European Union

Policies/Objectives cited	United Kingdom	United States	European Union
Agreements with vehicle manufacturers	• UK car manufacturers are committed to meeting the ACEA target of a 10% improvement in vehicle efficiency in 2005 (I); Government also supports Commission proposals for a negotiated agreement. • Government participates in Greener Motoring Forum which aims to co-ordinate voluntary initiatives between the public, private and voluntary sectors. Activities include "Tune Your Car" campaign and a publication containing environmental information on new cars (I).		(See Regulations and Guidelines above: Strategy element on CO_2 emissions from cars).
Alternative fuels		• "Clean Cities" Programme run by the US Dept. of Energy, designed to expand use of alternatives to petrol and diesel fuel. Partnerships with local businesses and Governments to promote viable alternative fuels market (I).	
Other		• "Transportation Partners" Programme: Run by US EPA designed to reduce growth in greenhouse gas emissions from the transportation sector by showing growth in vehicle miles travelled. Partnerships with non-governmental organisations, local governments, citizens' organisations, to promote policies/projects to reduce VMT (I).	

Table 7a. INFORMATION AND TRAINING INITIATIVES
Canada, Denmark, France, Japan

Policies/Objectives cited	Canada	Denmark	France	Japan
Information actions and driver education	- Publication and distribution of the annual **Fuel Consumption Guide** which informs purchasers of new cars, light trucks and vans about the fuel efficiency of these vehicles: part of Motor Vehicle Fuel Efficiency Initiative. (I: Guide distributed to more than 400 000 parties per year.) - Buy $mart programme, part of the Motor Vehicle Fuel Efficiency Initiative. A new labelling initiative that will provide consumers with fuel efficiency information on individual vehicles (A). - As part of the Alternative Transportation Fuel (ATF) Market Development Initiative, Natural Resources Canada is working with the ATF industry and vehicle manufacturers in Canada to promote alternative fuels, propane in particular. (I: Over 140 000 propane vehicles are on the road served by some 5 000 fuelling stations.) - Auto$mart programme: Private motorists purchasing new vehicles are provided with information on vehicle energy efficiency, maintenance and driving habits. (I: Since the 4th quarter of 1994, over 4 000 telephone enquiries were made to Auto$mart's hot-line). - Energy efficiency training under the Auto$mart programme for new drivers to help them understand and adopt practices to minimise fuel consumption and vehicle emissions (I). - Fleet Energy programmes: to encourage energy efficiency and greater use of alternative fuels in public sector and commercial fleets. • FleetWise: helps federal government fleets via information materials, workshops and training programmes (I: launched in 1995). • Fleet$mart: helps fleet operators in other sectors save fuel and reduce operating costs by recruiting fleets to participate in the programme; developing information products and training modules on fuel-efficient driving; and identifying remaining barriers to increasing fuel efficiency in fleets (A: anticipated launch in early 1997).	Training in energy-efficient and environmentally-friendly driving behaviour. Organisations and drivers' associations urged to conclude agreements for economical driving; additional course set up for current drivers (I).	Financial support from ADEME (state-backed energy-efficiency R&D group) for development of computer programme to be used for teaching economic driving (I: available for demonstration end 1994; aim is to equip 800 driving schools by 2000).	Mandatory fuel economy labelling subject to penalty (I: strengthened in 1993).

100

Table 7b. **INFORMATION AND TRAINING INITIATIVES**
Latvia, Lithuania, Netherlands, New Zealand

Policies/Objectives cited	Latvia	Lithuania	Netherlands	New Zealand
Information actions and driver education	Driver education on efficient use of vehicles, maintenance of vehicles in good technical condition with a view to decreasing fuel consumption and greenhouse gas emissions (I) (no official policy).	Improvement of driver selection and training (I).	NOVEM programme: "Buy Eco-wise, Drive Eco-nice". Action programme on fuel-efficient driver behaviour and vehicle purchasing practices, targeting company car drivers, truck and van drivers, commuters and private car drivers. Financing of F2 million per year from Transport, Environment and Energy ministries for information actions and driver education and training, as well as R&D on effects of econometers, cruise controls, etc. (I: since 1992) Fuel-efficiency labelling system of all new cars corresponding to the envisaged fiscal incentive scheme (P).	Driver education on speed limits (I: see Regulations and Guidelines above).

Table 7c. **INFORMATION AND TRAINING INITIATIVES**
Russian Federation, Slovak Republic, Switzerland, United Kingdom

Policies/Objectives cited	Russian Federation	Slovak Republic	Switzerland	United Kingdom
Information actions and driver education	Preparation of annual report on impact of transport on the environment (I).	Information, education campaigns aimed at new drivers; drivers training focusing on techniques and vehicle maintenance to encourage better fuel economy/minimisation of impacts on environment (status not communicated).	Information campaigns promoting urban traffic management, car-sharing, and driver education in the context of the Energy 2000 action plan's energy economy programme for transport (I).	"Tune Your Car" campaign intended to raise awareness of links between good car maintenance and environmental impacts; scheme to provide environmental information on new cars -- both co-ordinated by Greener Motoring Forum (see VA) (I). Department of Environment and Freight Transport Association published guide to fuel consumption in freight haulage fleets, offering practical guidance on how companies can improve efficiency (I).

Table 7d. INFORMATION AND TRAINING INITIATIVES
United States, European Union

Policies/Objectives cited	United States	European Union
Information actions and driver education	• Fuel Economy Labels for Tyres: The Climate Change Action Plan calls for a mandatory tyre fuel-efficiency labelling programme for new and replacement tyres for light duty vehicles. Voluntary programme for heavy-duty trucks and buses proposed by Department of Transportation. [P: Regulation (LVs)/testing to implement (HGVs) programmes pending]. • Fuel Economy Guide: US Department of Energy required to publish an annual guide for consumers of fuel economy and fuel cost estimates for each new vehicle available in the new model year. All US new car and light truck dealers required to have available and prominently display a copy of the guide (I: Since 1975). • Fuel Economy Labels for new cars: Consumer information fuel economy label required to be displayed on each new car and light truck sold in the US beginning in model year 1976. Current label provides fuel economy and fuel cost estimates for a given vehicle as well as the range of fuel economy ratings for other new vehicle models in the same class size (I: Since 1975).	(See Regulation and Guidelines above: Strategy element on fuel economy labelling).

Table 8a. RESEARCH AND DEVELOPMENT
Austria, Canada, Denmark, France

Policies/ Objectives cited	Austria	Canada	Denmark	France
Alternative-fuelled or powered vehicles / new technologies	Promotion of new technology (electric/ hybrid vehicles); greater use of new vehicle technologies (I: Part of Five Points Programme in progress).	Alternative Transportation Fuel R&D Programme: focuses on the development of competitive, energy-efficient and environmentally responsible technologies. (I: in 1996, three full-sized hydrogen fuel-cell buses were put in service in Vancouver).		
Environmentally-friendly transport means and equipment			Development of more energy-efficient transport means; improve vehicle fuel economy (status not communicated; Denmark to pursue this in international channels).	Under PREDIT programme (R&D for innovation and technology in land transport) FF 445 million allocated to high-speed railways 1990 to 1994, notably for new generation TGV (I).
Combined transport				Under PREDIT programme, from 1990 to 1994, FF 450 million devoted to inter-modal transport research (I: plans to continue and extend research).
Vehicle fuel economy				Under PREDIT programme, FF 1.2 billion invested from 1990 to 1994 for vehicle fuel consumption. EU-level initiative favoured (I).

Table 8b. RESEARCH AND DEVELOPMENT
Japan, Latvia, New Zealand, Poland

Policies/Objectives cited	Japan	Latvia	New Zealand	Poland
Alternative-fuelled or powered vehicles / new technologies	Improve technology to enhance performance and commercialisation possibilities including: alternatives to lead batteries; battery exchange system; element technologies (high-efficiency motors); CNG-fuelled, and hybrid-engine vehicles; methanol-fuelled and hydrogen-powered vehicles (basic research) (I).			Introduction of "modern technologies" (P) (also via EI and R/G).
Environmentally-friendly transport means and equipment		Scientific studies planned in this area with a view to reducing fuel consumption and greenhouse gas emissions (P).		
Economic instruments			Development of procedures for incorporating CO_2 into cost-benefit analysis (status not provided).	

Table 8c. **RESEARCH AND DEVELOPMENT**
Romania, Russian Federation, Sweden, Switzerland

Policies/Objectives cited	Romania	Russian Federation	Sweden	Switzerland
Alternative-fuelled or powered vehicles / new technologies	Increase in technical efficiency of automobiles in order to reduce new-vehicle fuel consumption via use of economic instruments and R&D (E).		SKr 240 million allocated to co-fund demonstration programme for electric/hybrid electric vehicles (I: 1994-1995); SKr 240 million in partial funding of projects to encourage alternative fuel use possibilities: trials have involved alcohol-, biogas-, or natural gas-powered buses; studies include: fuel-supply infrastructure, development of engines for heavy traffic optimised for a given fuel (I).	
Environmentally-friendly transport means and equipment				The National Research Programme no. 41 (NRP 41) was created as a think tank for sustainable transport policy. Among others, the NRP 41 is working on projects dealing with potential for walking and biking policies and demand projections for a Swiss metro (I).
Combined transport				As part of NRP 41 (see previous entry) a project is under way looking into potential and locations for combined transport (I).
Vehicle fuel economy		Production and use of more energy-efficient vehicles (E).		
Economic instruments		Development of economic instruments for use in response to emissions regulations (E).		As part of NRP 41 (see previous entry), project underway examining use of databases for traffic and mobility management (I).
Emissions monitoring		Improvements in ecological control and monitoring and greenhouse gas inventory systems; development of greenhouse gas source data system (E).		

Table 8d. **RESEARCH AND DEVELOPMENT**
United Kingdom, United States, European Union

Policies/ Objectives cited	United Kingdom	United States	European Union
Alternative-fuelled or powered vehicles / new technologies	The Government is a partner in a two-year trial to assess the viability of alternative fuels (including CNG, LPG, electricity and biodiesel), their cost and emissions performance in real operating conditions. Results will be published in 1997 (I: ongoing). Modelling work carried out for the Government suggests that alternative fuels are more costly and less practical than petrol and diesel but that there are opportunities in certain market segments (I).	Partnership for a New Generation of Vehicles. Joint effort of government and United States Council for Automotive Research. Goals: improve national competitiveness in manufacturing, implement commercially viable innovations from research; develop a vehicle of up to triple the fuel efficiency of today's vehicles (I: Initiated in 1993; recognized in the Climate Change Action Plan and elsewhere as a critical component of the US's long-term climate change strategy).	THERMIE Programme to promote new energy technologies through pilot and demonstration projects. In 1992: 21 transport-related projects (primarily urban transport); 5% of total project funding (I: Since 1978). JOULE Programme to promote research in improving energy efficiency. Energy efficiency improvements in internal combustion engines; battery-driven systems/ fuel-cell development in electric vehicles; transport modelling (I). BRITE Programme to promote research in industrial materials and clean manufacturing technologies, particularly advanced internal combustion engines/alternative propulsion technologies (hybrid vehicles, gas turbines, CNG) (I: since 1991). ALTENER Programme to promote renewable energy sources in the market. Aims to secure 5% market share of total motor vehicle fuel consumption for biofuels (I: since 1993).
Environmentally-friendly transport means and equipment			SAVE Programme : 13 pilot projects in the fields of transport and traffic management; two pilot studies receive total Community support of approx. ECU 2.1 million (I: since 1993).
Economic instruments	Government commissioned major study into congestion charging in London; research showed significant potential benefits but Government concluded that scheme could not be implemented this century (I: research ongoing). Programme to research, develop and test electronic technologies for motorway tolling (I: research in early stages).		

4.3 Summary and remarks

Summary

Table 9 below summarises how ECMT countries responded to the policies and measures section of the questionnaire. The numbers represent the number of times that types of measures and particular aspects of the table were answered. They are only indicative, as the information provided was not always specific enough to draw conclusions, and as many measures listed could be housed in several different categories.

Table 9. **Responses to Table 7 of the Questionnaire on Policies and Measures**

Country	Type and No. of Instrument / Approach[1]						Status reported[2]					Progress Indicators	Estimated Effect in 2000	Quantitative analysis	Emissions tables link
	R/G	EI	VA	IN	TR	R&D	I	A	P	E	O				
Austria	6	5	1			1	10	2		1		2		1	
Belgium	3	7					6	4				1	4		
Canada		3	1	5	2	1	9	3				11			
Czech Republic	7	7	4				12	6				3	4		
Denmark	3	6		1		1	7				4*		2		
Finland	5	11					11		2	2	1*	2	1		
France[3]	3	16	2		1	3	18	6		1					
Germany	2	8	1				10	1							
Hungary	5	5					9	1				6	1		3
Ireland		11					9	1	1						
Italy	1	16					10	5			2*		15		
Japan	3	10		1		1	15					1	1	1	
Latvia	5	3		1		1	6	4							
Lithuania	3	4	1		1		7			1	1*				
Netherlands	6	14	1	1	1		18	3	2				7		
New Zealand	4	1	1		1		4				3*	1			
Norway	3	5					7			1		1			
Poland	2	5			1			1	4	3				4	5
Portugal	5	4					9								
Romania		1			1					2		2	2	2	2
Russian Federation	6		1		3		2	1	2	5					
Slovak Republic	4	10		1			10	1		2	2*		9	3	10
Spain		5					3				2*				
Sweden	3	3	1		2		7	1	1						
Switzerland	5	3		1		3	11		1				1		
United Kingdom	4	5	2	2		4	16		1				1	2	1
United States	4	3	2	3	1		11	2					4	4	
European Union	2	1				5	6	1	1						
TOTAL	94	172	16	16	7	29	243	33	25	18	15	30	52	17	21

Source: 1996 ECMT Questionnaire.

Notes:

* Status not specified
1. **R/G**: Regulation/Guidelines; **EI**: Economic Instrument; **VA**: Voluntary Agreement/Action; **IN**: Information action; **TR**: Training; **R&D**: Research and Development.
2. **I**: Approved by Government and Implemented; **A**: Agreed by Government -- not yet implemented; **P**: Proposed in policy statement --agreement pending; **E**: Envisaged as potential measure -- no official policy proposal made; Other.
3. Information not provided in questionnaire; taken from national communication directly, numbers approximate.

The chart below shows the share of each policy category in the total 318 policies listed in the responses to the questionnaires or taken from the National Communications to the FCCC.

Table 10. **Share of Types of Policies and Measures in Total**

Type of Policy/Measure	Number of times cited	Percentage share in total (approx.)
Economic Instruments	172	51
Regulations and Guidelines	94	28
Research and Development	29	9
Voluntary Agreements/Actions	16	5
Information Actions	16	5
Training	7	2
TOTAL	**334**	**100**

The following six tables show which countries indicated use of or plans to implement specific measures to target CO_2. Please note that the absence of a particular country in a specific policy category does not necessarily mean that the country has not taken action in that area; it just means that the country did not list that action as a measure to limit CO_2.

Table 11a. **Policy categories by country**
Economic Instruments

Carbon or CO_2 tax	Denmark, Finland, Norway, Sweden.
Fuel taxes to discourage excessive use of road vehicles	Austria, Belgium, Denmark, Finland, France, Germany, Ireland, Italy, Lithuania, Netherlands, New Zealand, Poland, Slovak Republic, Sweden, Switzerland, United Kingdom.
Fuel tax exemption/reduction to promote use of certain fuels	Belgium, Finland, France, Latvia, Slovak Republic, United Kingdom.
Government subsidies/other support to encourage use of public transport/inter-modal transport/modal shifting/ discourage excessive road transport	Czech Republic, Denmark, Finland, France, Germany, Ireland, Italy, Japan, Latvia, Lithuania, Netherlands, Norway, Poland, Portugal, Romania, Slovak Republic, Spain, Switzerland, United Kingdom.
Use of subsidies for traffic management systems	France, Germany, Ireland, Italy, Japan, Lithuania, Netherlands, Portugal, Slovak Republic, Sweden.
Use of subsidies to encourage use of alternative-powered vehicles/new technologies	Canada, Czech Republic, Denmark, France, Hungary, Ireland, Italy, Japan, Netherlands, Norway, Poland, Spain, United States.
Road taxes and road pricing	Austria, France, Germany, Netherlands, Norway, Poland, Slovak Republic.
Fiscal measures to promote sale of more fuel-efficient vehicles	Austria, Canada, Finland, France, Germany, Ireland, Italy, Japan, Netherlands, Norway, Slovak Republic, United States.
Vehicle fleet renewal	France, Hungary, Italy, Slovak Republic, Spain.
Fiscal measures to encourage use of public transport for professional commuter traffic/reduce incentive for business travel	Belgium, Denmark, Finland, Ireland, Netherlands, United States.
Fiscal advantages for public transport companies	Belgium, France, United Kingdom.
Tariff adjustment to encourage public transport use	Czech Republic, Netherlands, Spain.
Actions to support combined transport	Austria, Belgium, Czech Republic, Switzerland.
Other	Hungary, Italy, Lithuania.

Table 11b. **Policy categories by country**
Regulations and Guidelines

Heightened speed limit enforcement to decrease fuel consumption; increase safety.	Austria, Belgium, France, Italy, Latvia, Netherlands, New Zealand, Norway, Portugal, Slovak Republic, Switzerland, European Union.
Mandatory vehicle inspections	Finland, France, Hungary, Latvia, Netherlands, Norway, Portugal, Slovak Republic, Switzerland, United States.
Traffic management; restrictive policies to discourage private vehicles in city centres/reduce fuel consumption	Austria, Belgium, Czech Republic, Latvia, Netherlands, Portugal, Slovak Republic, United Kingdom.
CO_2 emissions standards	Czech Republic, Russian Federation.
Vehicle fuel consumption standards	Czech Republic, Japan, Russia, Sweden, Switzerland, United States.
Development of programmes/guidelines on climate change/transport and environment	Austria, Czech Republic, Denmark, Finland, Hungary, Latvia, Lithuania, Netherlands, New Zealand, Norway, Poland, Portugal, Russia, Sweden, Switzerland, United Kingdom, United States, European Union.
Measures to promote public transport	Finland.
Measures to promote combined transport	Japan, Slovak Republic.
Mandatory transport plans for companies/ measures to promote telecommuting	Belgium, United Kingdom.
Efforts to integrate transport and land-use planning	Finland, Netherlands, New Zealand, Poland, Portugal, United Kingdom.
Import restrictions on highly polluting vehicles (especially in CEE countries)	Hungary.
Other	Austria, Czech Republic, Germany, Hungary, Lithuania, Switzerland, United States.

Table 11c. **Policy categories by country**
Voluntary Agreements and Actions

Agreements with vehicle manufacturers	Austria, Canada, Czech Republic, France, Germany, Sweden, United Kingdom, European Union.
Alternative fuels	Czech Republic, United States.
Combined transport	Czech Republic.
Other	Lithuania, Netherlands, United States.

Table 11d. **Policy categories by country**
Information and Training Initiatives

Driver education and information actions	Canada, Denmark, France, Japan, Latvia, Lithuania, Netherlands, New Zealand, Russian Federation, Slovak Republic, Switzerland, United Kingdom, United States, European Union.

Table 11e. **Policy categories by country**
Research and Development

Alternative-fuelled or powered vehicles/new technologies	Austria, Canada, Japan, Poland, Romania, Sweden, United Kingdom, United States, European Union.
Environmentally-friendly transport means and equipment	Denmark, France, Latvia, Switzerland, European Union.
Combined transport	France, Switzerland.
Vehicle fuel economy	France, Russian Federation.
Economic instruments	New Zealand, Russian Federation, Switzerland, United Kingdom.
Emissions monitoring	Russian Federation.

Remarks

A number of observations can be made when examining the way ECMT countries described their policies to limit CO_2 from transport in the responses to the questionnaire.

In general:

-- Countries that responded to the questionnaire cited a vast and various range of policies and measures in the area of transport and environment as elements of their plans to limit CO_2 from transport. Measures listed ranged from the specifically targeted « CO_2 tax » in four ECMT countries, to the significantly more broad-based « subsidies to improve public transport ». It appears clear that considerable efforts are being made by ECMT countries to build more sustainable transport systems and behaviour patterns: half of the countries responding noted initiatives to increase the price of petrol or diesel ; more than half cited government support to public/mass transport. A large share of these reported measures, though, only indirectly address the problem of CO_2. It seems, therefore, that despite their commitments to the FCCC, many ECMT countries are still in the early stages of developing specific strategies to limit CO_2 from transport.

-- Descriptions of some policies and measures were frequently insufficient to draw conclusions as to their impact and effect on CO_2. The questionnaire was designed to solicit the greatest degree of detail possible. In a number of cases, despite requests for further information, descriptions of the measures, their status, the time frame of their implementation, etc. were not complete enough to ascertain their role in CO_2 abatement; as a result, they were not included in the compilation of measures above. Attempts were made, nevertheless, to include as many of the measures cited as possible, even if the link with CO_2 was distant.

-- It will be essential for signatories of the FCCC in the coming months to be able to design robust policy packages to address climate change, notably the role of transport in the problem. Without clear, comprehensive descriptions of the policies, evaluation of their projected impacts on CO_2 will continue to be difficult. As noted earlier, lack of clarity and transparency in policy descriptions has also been evoked by the FCCC in its first review of national communications. The second round of communications, which will begin to be submitted in 1997, will no doubt show improvements in this area.

As concerns the policies and measures themselves:

-- It is evident from the questionnaire responses that <u>economic instruments</u> are by far the mechanism of choice for addressing CO_2 in ECMT countries, accounting for approximately 51 per cent of all measures cited. Of particular note :

⇒ 19 countries cited use of subsidies or other forms of government support to encourage public transport use/inter-modal transport/modal shifting ;

⇒ 16 countries mentioned use of fuel taxes as a means of increasing the price of certain fuels - primarily petrol – to restrain consumption ;

⇒ 12 countries noted fiscal measures to promote sale of more fuel-efficient vehicles;

⇒ Only four countries -- all Scandinavian -- have implemented carbon or CO_2 taxes.

-- Following Economic Instruments as the type of measure most-often cited was <u>Regulations and Guidelines</u> with approximately 28 per cent of the total. It is noteworthy that the three types of measures most often mentioned in this category, however, have more to do with promoting overall sustainable transport systems than with CO_2 abatement in particular.

⇒ 18 countries -- over 50 per cent of the responses to the questionnaire -- noted development of programmes or guidelines on climate change/ environment and transport. Only six of the programmes/guidelines reported as implemented or envisaged in this category (from Czech Republic, Denmark, Netherlands; Russia, Sweden, European Union) actually address climate change and transport specifically. The others deal mostly with larger transport and environment/sustainable transport issues.

Given that targets on climate change for the period beyond 2000 are currently being negotiated in the FCCC process, countries should perhaps now be exploring how programmes in place or being considered can work more effectively to meet the requirements to come.

⇒ 12 countries mentioned tighter speed limit enforcement as a way to reduce fuel consumption ; the principal impetus behind speed limit control was in most cases, however, safety and not lower consumption and emissions.

⇒ Ten countries cited mandatory vehicle inspections, eight noted traffic management, six cited vehicle fuel consumption standards as aspects of their CO_2 abatement plans, and six mentioned integration of land use and transport planning

Inasmuch as traffic management, mandatory vehicle inspections, and integration of land use and transport planning contribute to fewer vehicle kilometres travelled on the roads, and therefore less fuel consumed, they are, in effect, potentially significant in CO_2 abatement. They do not, however, directly address CO_2 as other actions in this category might, such as standards for CO_2 emissions standards or vehicle fuel efficiency. Two countries noted plans to study the option of CO_2 emissions standards.

-- Measures in the area of Research and Development accounted for approximately 9 per cent of all actions cited, the third most-frequently noted category, though significantly less than Regulations and Guidelines. Nine countries noted Research and Development in progress or under consideration in alternative-fuelled/powered vehicles ; five in the more general area of environmentally-friendly transport means and equipment ; and four in economic instruments.

The demonstration aspect seems to be a focal point in many research and development initiatives.

Comparatively few Voluntary Agreements/Actions were mentioned, representing only about 5 per cent of all measures cited.

Bearing particular mention are current national initiatives to conclude agreements on fuel efficiency targets with domestic vehicle manufacturers. The German automobile industry has committed to reducing average CO_2 emissions from their cars by 25 per cent from 1990 to 2005. French car manufacturers have set a target of cutting average CO_2 emissions from their vehicles to 150g per km by 2005. And Swedish manufacturer Volvo has stated that it will lower average fuel consumption of its cars sold in the European Union by 25 per cent from 1990 to 2005. Austria is also studying the feasibility of a similar kind of accord with industry.

France and the United Kingdom have both set up government-industry groups to address particular aspects of transport and environment issues.

A number of countries classified measures as Voluntary Actions with little to no explanation of what the voluntary aspect of the policy was or who the participants in the agreement were; consequently, it was difficult to include them in the policy matrices above.

-- Information and training initiatives were also few in number, accounting for around 5 per cent and 2 per cent of total measures respectively.

⇒ Canada, the Slovak Republic, Switzerland and the United Kingdom listed information campaigns targeting drivers.

⇒ Efforts to provide training in more fuel-efficient/environmentally friendly ways of driving were cited by Canada, Denmark, Latvia, Lithuania, the Netherlands, and New Zealand.

⇒ Fuel economy labelling is mandatory in Japan and the United States, and proposed for tyres in the United States. The European Union is considering fuel economy labelling as a principal part of its three-prong strategy to address CO_2.

As concerns the way the questionnaires were completed:

Countries were requested to denote where possible the *status* of the measures listed, *progress indicators*, the *estimated effect* of the measures in 2000, if *quantitative analysis* had been carried out on the effect of the measures in reducing CO_2, and if a *link with the emissions data* could be made.

Status

As regards the recording of status for the measures: understanding in what time frame a measure is implemented is essential to evaluate its impact on emissions. In numerous questionnaire responses, the status of a particular measure was not provided or could not be inferred from the information given.

As shown in Table 9, respondents to the questionnaire primarily reported measures already implemented (243) or Agreed, although those Agreed were cited significantly less frequently (33). Relatively few measures were listed as proposed or envisaged (25 and 18 times respectively). And status was unable to be determined in 15 cases.

Progress Indicators

Indicators of progress were reported for only 30 out of the 334 measures. Moreover, where mentioned, progress indicators were not always linked to the policy itself but instead to larger circumstances, or even actions carried out by international bodies of which a particular country was a part.

This may partially reflect that either monitoring of the effects of individual measures is insufficient, or that results of particular policies have not been quantified on a measure-by-measure basis.

Estimated effect in 2000

Recordings of estimated effects of policies in 2000 were relatively few -- around 16 per cent of all measures noted -- and diverse in their clarity and relevance. Degree of specificity ranged from the very general « air pollution in big cities will decrease » to the more precise « transport sector savings of 13 PJ or 25 per cent of total energy savings ». In most cases, information was insufficient to understand the rationale behind the estimations. It would appear that ability to more clearly determine the effect of policies will improve as strategies for CO_2 abatement are more precisely defined and emissions accounting systems are refined.

Quantitative analysis

Quantitative analysis was noted as having been carried out for 17 of the 334 measures. As was the case in the previous section, responses for quantitative analysis ranged from detailed descriptions of modelling exercises undertaken to the more succinct " yes " or " no ". While many countries seem to be in the process of improving quantitative evaluation of policies and measures, the few recordings of quantitative analysis in the questionnaire render difficult the drawing of any meaningful conclusions with regard this aspect.

Link with emissions tables

One of the objectives of the questionnaire was to get an idea about how policies and measures discussed in ECMT's interim report on CO_2 emissions were -- if at all -- reflected in emissions inventories and projections three years later.

Out of the 29 responses to the questionnaire, five countries noted links between policies and the emissions tables. Even when the numbers of the tables were noted, however, the impact of individual measures was difficult to ascertain, as little explanation as to how the links were made, or what the share in the impact on emissions was for a given policy or measure, was provided.

Based on the responses to the survey, it can be concluded that the link between policies reported and their effect on currently available emissions inventories is not yet being made on a widespread basis.

When one considers that:

-- initial commitments to address the climate change problem were only taken in 1992 in many countries, and that ratification of those commitments dates only to this year in several countries;
-- the scientific debate has only borne preliminary conclusions in the last year;

it is not incomprehensible that most ECMT countries have not yet been able to produce clear-cutting, quantifiable evidence of the effectiveness of certain policies in reducing emissions of CO_2. Indeed the time necessary to bolster political impetus to effect change has no doubt been insufficient. This being said, countries appear to be working toward development of more sustainable transport systems and behaviour patterns. The time is therefore propitious to capitalise on these very positive initiatives and make them work more effectively toward the more specific goal of limiting the share of transport in climate change.

5. CONCLUSIONS

The information provided in the ECMT survey reveals that few countries to date have developed focused strategies for CO_2 abatement from transport; the majority of countries has not yet put such strategies in place. This is not too surprising, given the status of the international climate change agenda at this time: as noted earlier, international perspectives and agreements on climate change are still in gestation, and the FCCC has only recently been ratified in some ECMT countries.

The international agenda is rapidly moving forward, however, and new, binding targets are being negotiated, even though initial emissions targets -- to which a majority of ECMT countries have ascribed -- will be met in only very few cases. The emissions data provided by ECMT countries in their responses to the questionnaire show that targeted stabilisation and reduction of CO_2 will by and large not be achieved in the time frame set forth when the FCCC was signed. Moreover, projected emissions in general, and in particular from the transport sector, are rising in significant proportions in most countries.

When comparing the CO_2 data trends revealed in the survey with the policies and measures that ECMT countries have put into action or have stated their intention to implement, it is clear that more focused, concerted policy development must take place in this area. CO_2 emissions from transport -- road transport in particular -- are on the rise in both relative and absolute terms and projected to continue on that trajectory through 2010, in spite of policies and measures designed to limit CO_2. This is not a revelation; it has been iterated and reiterated manifold times in ECMT and other fora. The question remains, however, what to do. On the one hand, looking beyond the transport sector itself, the scientific debate continues as regards the fundamental issue of how -- if indeed at all, and if so, in what time frame -- anthropogenic activity impacts climate change. This element of uncertainty may be a disincentive to some governments to take actions that are perceived as potentially costly to industry as long as the science is not fully conclusive. On the other hand, the International Panel on Climate Change, a large body of scientists from around the world, has stated that there is sufficient reason to believe in the link between human activity -- including transport -- and potential climate change. Based on this, the FCCC process -- of which virtually all ECMT countries are participant -- has determined that political action is warranted, and short-term binding targets to limit greenhouse gases -- foremost among them CO_2 -- appropriate for developed countries.

ECMT countries have made initial strides toward addressing CO_2 from transport. Initiatives are underway to render databases more disaggregated and more sophisticated in terms of the information they produce; countries continue to seek more effective policies to limit CO_2; and several countries have already designed specific strategies to minimise transport's role in climate change. There seems, however, to be a long way yet to go in order to bring about the stabilisation and reduction in emissions sought. A number of observations and recommendations can be made in light of the information examined in the body of this report.

As ECMT countries pursue ways to better address transport-related CO_2 emissions, it is recommended that a more strategic, targeted approach to policy development be considered; this will help to avoid the citing of a wide and diverse range of policies, many carrying little indication of their potential impact on emissions. A more-focused, well-developed plan to address the problem will go a long way to improving efforts to monitor the specific effects of policies on CO_2; likewise, better monitoring of the impact of certain policies will enhance transparency and efficiency in policy development and provide useful lessons for others.

In adopting a more targeted approach to transport and CO_2, countries might consider designing a package of cost-effective policies and measures as an integrated part of economy-wide actions to reduce greenhouse gases. Policy packages should address the variety of technological, regulatory, and economic aspects of the problem, sending the right messages to transport system users, system providers and vehicle manufacturers. Voluntary initiatives should be promoted on the part of industry where not yet under way, and dialogue strengthened where applicable between the policy-making bodies and the automobile industry, especially in large vehicle markets. Policy packages have the benefit of offering a comprehensive approach to a given problem; this can help to avoid the selection of individual policies that when implemented in isolation can lead to undesired countervailing effects.

In the short term, countries should seek and seize opportunities to implement cost-effective or "win-win" measures that can be introduced early on and lead to additional benefits -- such as local air pollution abatement, better road safety, and greater overall efficiency in the transport sector -- while also limiting CO_2 emissions. Examples of such measures might include: tighter enforcement of speed limits; more stringent vehicle inspection systems; information campaigns and education to improve driver behaviour; efficient structures of fuel and vehicle taxation; better fleet management; and vehicle loading factors.

Communication of policy actions undertaken or envisaged is a big part of FCCC requirements, and on a more-limited scale, of the ECMT survey as well. Better communication of these initiatives can lead to improved understanding of their benefits and disadvantages, which in turn can produce more effective policy development. The responses to the survey demonstrate this as an area where improvement is needed among ECMT countries.

Much remains to be fully understood as to how different policies work individually and in combination to reduce CO_2 emissions. Efforts to improve transparency can bring about better understanding in this area. With a more strategic policy approach to the transport-related CO_2 problem will most likely come enhanced clarity as to how to solve it.

Suggestions for further work to be done

Given the short-term targets and deadlines now established by the international climate change agenda, and the important role transport plays in the climate change phenomenon, it is suggested that it would be useful to:

-- Repeat this exercise in about four years

The second round of National Communications to the FCCC will begin to be submitted in early 1997. Further, COP-3 will take place in December 1997, at which quantified limitation and reduction objectives will be set within specified time-frames for developed countries. Developed countries which have signed the FCCC will be obliged in the future to meet these targets. Therefore, adopting effective and efficient strategies and measures to limit transport's contribution to the climate change problem will be increasingly necessary. The periodic monitoring of how ECMT's Members and Associate Members are addressing this issue can be useful in evaluating progress towards these objectives.

-- Share this experience with the larger international forum on climate change

Virtually all ECMT Member and Associate Member countries have a stake in the outcome of the international climate change negotiations, having set stabilisation and reduction targets in the context of the FCCC process. Transport is an important part of that debate. ECMT can support the on-going discussions by making its experience in this area available to those directly engaged in this process, specifically, perhaps, by presenting these conclusions in advance of COP-3 later this year.

-- Provide targeted analysis of effective policies

Research on how certain policies and measures work to limit transport's role in climate change continues. Many questions remain as to the impact of policies under different circumstances. ECMT could contribute to this body of research, building on the questions brought forth in this report, and focusing on specific policy issues most pertinent to ECMT countries. An attempt to better understand barriers to implementation might also be useful to policy development within ECMT countries.

-- Explore ways to improve data base development and data monitoring

Sharing experience on the methodological aspects of data base development might be one way to improve overall tracking of CO_2 emissions, thus improving the capacity for and potential outcomes of comparative data analysis. ECMT could design an initiative to draw from the experience of its Members and Associate Members to share successes and failures.

NOTES

1. Signatories of the United Nations Framework Convention on Climate Change (FCCC) from industrialised countries agree to stabilise CO_2 emissions and other greenhouse gases at 1990 levels by 2000. There is flexibility in the Convention regarding the base year used by countries with economies in transition.
2. ECMT's new Members -- F.Y.R.O.M., Belarus and Ukraine -- were not yet full Members at the time the 1996 questionnaire was sent. For the purposes of this report, the term "ECMT countries" will be used to describe both Member and Associate Member countries and the European Union.
3. The existing target for industrialised countries (stabilization of greenhouse gas emissions at 1990 levels by 2000) is voluntary.
4. The United States specified that the targets should be binding, but not the individual measures, "thus allowing maximum flexibility in implementation" (Wirth, T., 1996).
5. In the context of the Framework Convention on Climate Change, the European Union is referred to as the European Community, the legal name deposited to the United Nations in 1994. The EU includes Austria (since January 1995), Belgium, Denmark, Finland (since January 1995), France, Germany, Greece, Ireland, Italy, Luxembourg, the Netherlands, Portugal, Spain, Sweden (since January 1995), the United Kingdom.
6. 13.2 ltr./100 km.
7. 8.6 ltr./100 km.

ANNEXES

ANNEX 1

QUESTIONNAIRE

In May and June 1996, ECMT member and associate member countries received the following questionnaire with an accompanying letter and instructions from the Deputy Secretary General.

QUESTIONNAIRE

Carbon Dioxide Emissions From The Transport Sector

1. Information Requested on Emissions

1.1 <u>Base-year data</u>

Please provide estimated emissions of CO_2 in thousand tonnes for the base year as used in the context of the Framework Convention on Climate Change. (For most signatory countries, this year will be 1990, however, the FCCC allows for some flexibility on this as concerns countries with economies in transition.) Please note that these data should already be available in your national government in the form of a greenhouse gas inventory, which is part of the "national communications" submitted to the FCCC by all parties.

<u>NOTE</u>: For all emissions data: please provide complete information on:

- sources of the data
- assumptions used
- comments on reliability of the data
- assumptions with respect to macroeconomic factors (e.g., growth in GNP) and microeconomic factors (e.g. number of cars, fuel economy).

The emissions include those, from all vehicles, vessels or aircraft, deriving from fuel sold on your territory.

TABLE 1. BASE YEAR EMISSIONS OF CO_2 (YEAR:)

SECTOR	SUB-SECTOR	TOTAL (000 TONNES CO_2)
ROAD TRANSPORT	Private Cars	
	Heavy Goods Vehicles and Buses	
	Motorcycles and Other[1]	
	Light Commercial Vehicles[2]	
TOTAL ROAD		
TOTAL RAIL		
SHIPPING	Internal Navigation	
	International Marine Bunkers[3]	
TOTAL SHIPPING		
AVIATION	Domestic Aviation	
	International Aviation Bunkers[3]	
TOTAL AVIATION		
TOTAL OTHERS (specify)		
TOTAL TRANSPORT SECTOR EMISSIONS		
TRANSPORT SECTOR EMISSIONS AS A SHARE OF TOTAL EMISSIONS		

Notes:

1. Indicate whether agricultural/ military transport emissions are included.
2. Specify if included in one of the other categories.
3. Although bunker fuel emissions are not included in *Total* emissions in FCCC context, please provide if available, and indicate approach used in accounting.

1.2 Projected CO_2 emissions data: 2000, 2010: Reference Case

Please provide below the emissions data as projected for years 2000 and 2010 (if available) in a Reference Case scenario (with no additional measures after base year). Please specify year in which modelling exercise undertaken.

TABLE 2. PROJECTED EMISSIONS OF CO_2: 2000, 2010: REFERENCE CASE
(with no additional measures after base year)

SECTOR	SUB-SECTOR	BASE YEAR (000 tonnes CO_2)	TOTAL 2000 (000 tonnes CO_2)	TOTAL 2010 (000 tonnes CO_2)
ROAD TRANSPORT	Private Cars			
	Heavy Goods Vehicles and Buses			
	Motorcycles and Other[1]			
	Light Commercial Vehicles[2]			
TOTAL ROAD				
TOTAL RAIL				
SHIPPING	Internal Navigation			
	International Marine Bunkers[3]			
TOTAL SHIPPING				
AVIATION	Domestic Aviation			
	International Aviation Bunkers[3]			
TOTAL AVIATION				
TOTAL OTHERS (specify)				
TOTAL TRANSPORT SECTOR EMISSIONS				
TRANSPORT SECTOR EMISSIONS AS A SHARE OF TOTAL EMISSIONS				

Notes:

1. Indicate whether agricultural/military transport emissions are included.
2. Specify if included in one of the other categories.
3. Although bunker fuel emissions are not included in *Total* emissions in FCCC context, please provide if available, and indicate approach used in accounting.
4. Indicate year in which modelling exercise undertaken.

1.3 Projected CO_2 emissions data: 2000, 2010: Status Quo Case

Please provide below emissions data as projected for years 2000 and 2010 (if available) for a Status Quo scenario, which should be based on the assumption that only measures <u>currently implemented</u> are applied in the whole period. Please note: <u>We understand that this data is not requested in the context of the FCCC; however, we would appreciate your including it if it is available.</u>

TABLE 3. PROJECTED EMISSIONS OF CO_2: 2000, 2010: STATUS QUO CASE
(with only measures already implemented since base year)

SECTOR	SUB-SECTOR	BASE YEAR (000 tonnes CO_2)	TOTAL 2000 (000 tonnes CO_2)	TOTAL 2010 (000 tonnes CO_2)
ROAD TRANSPORT	Private Cars			
	Heavy Goods Vehicles and Buses			
	Motorcycles and Other [1]			
	Light Commercial Vehicles [2]			
TOTAL ROAD				
TOTAL RAIL				
SHIPPING	Internal Navigation			
	International Marine Bunkers [3]			
TOTAL SHIPPING				
AVIATION	Domestic Aviation			
	International Aviation Bunkers [3]			
TOTAL AVIATION				
TOTAL OTHERS (specify)				
TOTAL TRANSPORT SECTOR EMISSIONS				
TRANSPORT SECTOR EMISSIONS AS A SHARE OF TOTAL EMISSIONS				

Notes:

1. Indicate whether agricultural/ military transport emissions are included.
2. Specify if included in one of the other categories.
3. Although bunker fuel emissions are not included in *Total* emissions in FCCC context, please provide if available, and indicate approach used in accounting.
4. Indicate year in which modelling exercise undertaken.

1.4 Projected CO_2 emissions data: 2000, 2010: Future Measures Case

Please provide below the CO_2 emissions data as projected for years 2000 and 2010 (if available), accounting for the effects of future measures to limit CO_2 (with measures that are already implemented or planned for implementation).

TABLE 4. PROJECTED EMISSIONS OF CO_2: 2000, 2010: FUTURE MEASURES CASE
(with measures that are already implemented or planned for implementation)[1]

SECTOR	SUB-SECTOR	BASE YEAR (000 tonnes CO_2)	TOTAL 2000 (000 tonnes CO_2)	TOTAL 2010 (000 tonnes CO_2)
ROAD TRANSPORT	Private Cars			
	Heavy Goods Vehicles & Buses			
	Motorcycles and Other[2]			
	Light Commercial Vehicles[3]			
TOTAL ROAD				
TOTAL RAIL				
SHIPPING	Internal Navigation			
	International Marine Bunkers[4]			
TOTAL SHIPPING				
AVIATION	Domestic Aviation			
	International Aviation Bunkers[4]			
TOTAL AVIATION				
TOTAL OTHERS (specify)				
TOTAL TRANSPORT SECTOR EMISSIONS				
TRANSPORT SECTOR EMISSIONS AS A SHARE OF TOTAL EMISSIONS				

Notes:

1. Include only measures that are "Implemented" or "Agreed" (see section 2 for definitions), and specify their status as such.
2. Indicate whether agricultural/military transport emissions are included.
3. If not separated, please indicate in which group included.
4. Although bunker fuel emissions are not included in *Total* emissions in FCCC context, please provide if available, and indicate approach used in accounting.
5. Indicate year in which modelling exercise undertaken.

1.5 CO$_2$ emissions data from the Road Transport subsector by fuel:

Please provide below CO$_2$ emissions data as projected for years 2000 and 2010 (if available) for fuel emissions from the Road Transport sector.

TABLE 5. EMISSIONS OF CO$_2$: 2000, 2010: FROM ROAD TRANSPORT (by fuel)

TYPE OF FUEL		TOTAL 2000 (000 tonnes CO$_2$)			TOTAL 2010 (000 tonnes CO$_2$)		
		Reference Case (with no additional measures after base year)	**Status Quo Case** (with only measures implemented since base year)	**Future Measures Case** (with planned measures)[1]	**Reference Case** (with no additional measures after base year)	**Status Quo Case** (with only measures implemented since base year)	**Future Measures Case** (with planned measures)[1]
Petrol							
Diesel	Heavy Goods Vehicles						
	Light Vehicles						
Other (Specify)							

Notes:

1. Include only measures that are "Implemented" or "Agreed" (see section 2 for definitions), and specify their status as such.

1.6 Annual emissions data from sector and by fuel

Please provide annual emissions data from the sector, broken down by subsector to the extent possible, and by fuel for 1990 to 1994 and 1995 if available. Annex 1 parties are requested to submit this information as supporting data -- though not at this level of disaggregation -- to the FCCC in early 1997; a copy of this data will be satisfactory.

TABLE 6a. ANNUAL CO_2 EMISSIONS DATA FOR SECTOR

SECTOR	1990 or base year	1991	1992	1993	1994	1995
ROAD TRANSPORT						
Private Cars						
Heavy Goods Vehicles & Buses						
Motorcycles and Other[1]						
Light Commercial Vehicles[2]						
TOTAL ROAD						
TOTAL RAIL						
SHIPPING						
Internal Navigation						
International Marine Bunkers[3]						
TOTAL SHIPPING						
AVIATION						
International Aviation Bunkers[3]						
Domestic Aviation						
TOTAL AVIATION						
TOTAL OTHERS (specify)						
TOTAL TRANSPORT SECTOR EMISSIONS						
TRANSPORT SECTOR EMISSIONS AS A SHARE OF TOTAL EMISSIONS						

Notes:
1. Indicate whether agricultural/ military transport emissions are included.
2. Specify if included in one of the other categories.
3. Although bunker fuel emissions are not included in *Total* emissions in FCCC context, please provide if available, and indicate approach used in accounting.

TABLE 6b. ANNUAL CO_2 EMISSIONS DATA BY FUEL

(in 000 tonnes CO_2)

TYPE OF FUEL		1990 or base year	1991	1992	1993	1994	1995
Petrol							
Diesel	Heavy Goods Vehicles						
	Light Vehicles						
Other (Specify)							

2. **Information Requested on Measures**

The attached summary table of policies and measures (Table 7) has been prepared for your country based on the national communication on climate change that your country provided to the Framework Convention on Climate Change and other sources. Using the table as a reference, please update, modify, and provide new information on all policies and measures in place or envisaged to limit CO_2 emissions from the transport sector.

Note: Clearly state the Instrument/ Approach and Status of all measures with the following abbreviations:

Instrument/ Approach: **R/G** = Regulation/Guidelines; **EI** = Economic Instrument; **VA** = Voluntary Agreement/Accord; **IN** = Information initiative; **TR** = Training; **R&D** = Research and Development.

Status: **I** = Approved by Government and Implemented; **A** = Agreed by Government -- not yet implemented; **P** = Proposed in policy statement -- agreement pending; **E** = Envisaged as potential measure; no official policy proposal made; **O** = Other (please specify). Provide as complete information on each measure as possible. As shown in the Table 6, please structure information in the following way:

In column:
1. a description of the policy/measure, including not only national measures, but also any relevant regional or local initiatives that are significant;
2. the objective of the measure (what it is expected to achieve);
3. the type of instrument or approach (see above paragraph);
4. the status of the measure at present (see above paragraph);
5. indicators of progress toward stated objectives (include description of how the measure is being/will be implemented, monitored and enforced);
6. the expected effects of the measure in 2000;
7. if quantitative analysis has been used to calculate emissions savings (yes or no) and, if yes, any key assumptions made in the calculations;
8. a cross-reference to Tables 2, 3 and/ or 4, to indicate to which emissions case the measure applies.

3. **Other information**

In order to understand as completely as possible the data provided, the measures described and their effects, we would appreciate your including the following background documents:

- Recent national policy statements or discussion documents on CO_2 from the transport sector;
- Plans or measures being discussed, but on which no decisions have been taken;
- Documents describing the cost-effectiveness of measures or analyses on performance of specific measures or programmes.

4. Contact person

For any questions regarding this questionnaire, please contact <u>Mary Crass at ECMT</u>:
<u>Tel</u>: +33 1 45 24 13 24 or 97 18, <u>Fax</u>: +331 45 24 97 42. Kindly provide below the name and contact numbers of the person with whom ECMT can follow up for questions and further information.

    Name:
    Position:
    Ministry/Department:
    Tel: Fax:

Note: Table 7 not included.

ANNEX 2

THE FRAMEWORK CONVENTION ON CLIMATE CHANGE: STATUS OF ECMT COUNTRIES

ECMT member countries have in great majority signed and ratified the United Nations Framework on Climate Change (FCCC), signed in Rio de Janeiro in 1992. To date, 29 of the 34 ECMT member countries and the European Union and all six ECMT associate member countries have ratified the Convention. The table below shows the status as of June 1996 of ECMT countries relating to the signing and ratification of the FCCC.

COUNTRY	FCCC STATUS	COUNTRY	FCCC STATUS
Member Countries		*Associate Member Countries*	
Austria	Ratified February 1994	Australia	Ratified December 1992
Belarus	N.A.	Canada	Ratified December 1992
Belgium	Ratified January 1996	Japan	Ratified May 1993
Bosnia Herzegovina	Has not signed the FCCC	New Zealand	Ratified September 1993
Bulgaria	Ratified May 1995	Russian Federation	Ratified December 1994
Croatia	Ratified March 1996	United States	Ratified October 1992
Czech Republic	Ratified October 1993		
Denmark	Ratified December 1993	*Observer Countries*	
Estonia	Ratified July 1994	Albania	Ratified October 1994
Finland	Ratified May 1994	Armenia	Ratified May 1993
France	Ratified March 1994	Azerbaijan	Ratified May 1995
Former Yugoslav Republic of Macedonia (FYROM)	Has not signed the FCCC	Georgia	Ratified July 1994
		Morocco	Ratified December 1995
Germany	Ratified December 1993		
Greece	Ratified August 1994	*Organisations*	
Hungary	Ratified February 1994	European Union	Ratified December 1993
Ireland	Ratified April 1994		
Italy	Ratified April 1994		
Latvia	Ratified February 1995		
Lithuania	Ratified March 1995		
Luxembourg	Ratified May 1994		
Moldova	Ratified June 1995		
Netherlands	Ratified December 1993		
Norway	Ratified July 1993		
Poland	Ratified July 1994		
Portugal	Ratified December 1993		
Romania	Ratified June 1994		
Slovak Republic	Ratified August 1994		
Slovenia	Ratified December 1995		
Spain	Ratified December 1993		
Sweden	Ratified June 1993		
Switzerland	Ratified December 1993		
Turkey	Has not signed the FCCC		
Ukraine	N.A.		
United Kingdom	Ratified December 1993		

ANNEX 3

COMPARATIVE DATA TABLES

Following are tables housing data on CO_2 emissions from transport received in response to the 1996 questionnaire. The tables are structured so that comparisons from country to country can be made, considering the different contexts for each country described in Section 3.1. Given the methodological and data accounting differences, the tables should be read with careful attention to the explanatory notes for each country following each set of data. The data is organised in the following sets of tables:

- Base-year Data
- Projected Emissions of CO_2 Reference Case
- Projected Emissions of CO_2: Status Quo Case
- Projected Emissions of CO_2: Future Measures Case
- Projected Emissions of CO_2 from Road Transport by Fuel
- Annual Emissions Data from Transport 1990-1995

Notes for all tables:

* Numbers have been rounded to nearest tenth in all tables where appropriate.
* "--" denotes «Not Available».
* Total emissions figures, when not provided, have been derived from the transport sector emissions categories and per cent share of transport in total.
* Refer to sub-sector tables for specific notes on emissions.

Table 12. Base-Year Emissions of CO_2 by Sector

(1990 = base year except where indicated)

	TOTAL EMISSIONS (million tonnes)	TRANSPORT EMISSIONS BY SECTOR (thousand tonnes)					TRANSPORT EMISSIONS (million tonnes)	PER CENT TRANSPORT
		Road	Rail	Shipping	Air	Others		
Austria	59.6	13 280[1]	707[2]	41[3]	1 002[4]	1 131[5]	16.16	27.1
Belgium	110	21 000		1 000[1]	10 300	--	22.0	20.0
Canada	451.9	117 800	6 300	5 700	10 300	--	140.1	31.0
Czech Republic	168.6	6 840[1]	738	54[2]	294[3]	--	7.9	4.7
Denmark[1]	60.2	9 239	483	445[2]	67[3]	--	10.2	17.0
Finland	65.0	11 200	200	1 600	520	2 100[1]	15.6	24.0
France[1]	382.7	111 500	1 100	8 000[2]	12 300	--	132.8	34.7
Germany	1 012	150 000	3 000	2 000	3 000	--	159.0	15.7
Hungary	75.1	8 131.6[1]	495.3	26.8[2]	532.3[3]	1 653.5[4]	10.3	13.8
Ireland	31.9	4 715	114	117	1 111	--	6.1	19.0
Italy	400.7	91 200	600[1]	9 600	5 800	2 600	109.8	27.4
Japan	1 174.8	--	--	17 800[1]	13 200[2]	--	215.0	18.3
Latvia	23.6	3 826	881.4	955[1]	168[2]	--	5.6	24.0
Lithuania	36.6	3 681	242	133	442	(1 293)[1]	4.5 (5.8)	12.3
Netherlands	184[1]	23 800	100[2]	1 500	500	1 800[3]	27.7	15.0
New Zealand[1]	27.9	--[2]	--	1 046	2 156	--	11.1	40.0
Norway	32.4	8 000	100	2 000[1]	1 300[2]	--	11.4	35.2
Poland[1]	--	20 016	7 328	588	566	--	28.5	--
Portugal	44.4	9 413	174	2 076	1 876	--	13.54	30.5
Romania[1]	--	6 832.2[2]	--	--	--	--	--	--
Russian Fed.	2 443.7[1]	146 900	1 840	18 100[2]	52 100[2]	--	234.6	9.6
Slovak Republic	53	4 500.8	376.9	142.6[1]	275.9[2]	--	5.3	10.0
Slovenia	13.6	2 947	62	--	--	183[1]	3.2	23.4
Spain	--	48 706	574	12 076	5 948	7 028	74.3	--
Sweden	60	16 100	100	2 800	1 500	--	20.5	34.2
Switzerland[1]	44.2	12 620	30	75[2]	3 544	975[3]	14.7	33.4
United Kingdom	575	109 691	1 933[1]	5 506[2]	2 614[3]	--	119.7	21.0

Source: Responses to 1996 ECMT questionnaire.

See notes after Table 13.

(thousand tonnes)

Table 13. **Base Year Emissions of CO_2 by Sub-sector**
(1990 = base year except where indicated)

	ROAD					SHIPPING			AVIATION		
	Private Cars	Heavy Goods Vehicles and Buses	Motorcycles and Other	Light Commercial Vehicles	TOTAL	Internal Navigation	International Marine Bunkers	TOTAL	Domestic Aviation	International Aviation Bunkers	TOTAL
Austria	8 471[1&6]	4 710	99	[6]	13 280	41[7]	--	41	1 002	--	1 002
Belgium	13 000	8 000	--	--	21 000		1 000[1]				
Canada	52 200	23 200	20 100[1]	22 300	117 800	5 700	--	5 700	10 300	--	10 300
Czech Republic	3 797		3 043[1]		6 840	54	--	54	294	--	294
Denmark[1]	5 013	2 089	--	2 137	9 239	445	--	445	67	--	67
Finland	6 850	3 200	--	1 150	11 200	--	--	1 600[2]	220[3]	--	520
France[1]	64 600	45 200	1 700		111 500	--	--	8 000[2]	--	--	12 300
Germany	110 800	38 500[1]	900	[1]	150 000	2 000	8 000[2]	2 000	3 000	11 000[2]	3 000
Hungary	4 202.5	3 299.6	45.6[1]	583.9	8 131.6	26.8	61	26.8	0	--	532.3
Ireland	--	--	--	--	4 715	56	--	117	--	1 111	1 111
Italy	53 700	35 000	2 600	--	91 200	1 200	8 400	9 600	1 800	4 000	5 800
Japan	--	--	--	--	--	--	17 800	17 800	--	13 200	--
Latvia	1 869	1 903	54	--	3 826	955	--	955	--	168	168
Lithuania	1 047	1 830	694	110	3 681	13	120	133	8	434	442
Netherlands	15 200	6 000	500	2 100	23 800	--	--	1 500	--	--	500
New Zealand[1]	--	--	--[2]	--[3]	--	--	1 046	1 046	789	1 367	2 156
Norway	5 800[3]	2 100	--	--	8 000	2 000[1]	--	2 000	--	--	1 300[2]
Poland[1]	10 796	4 756	--	4 464	20 016	--	--	588	--	--	566
Portugal	4 455		4 958[1]	[3]	9 413	889	1 187	2 076	946	930	1 876
Romania[1]	2 979	3 780.7[3]	72.5[2]	--	6 832.2	--	--	--	--	--	--
Russian Fed.	12 000	111 700	--	23 200	146 900	--	--	18 100	--	--	52 100
Slovak Republic	1 138.4	3 049.4[3]	47.1[4]	265.9	4 500.8	142.6	--	142.6[1]	--	--	275.9[2]
Slovenia	--	--	--	--	2 947	--	--	--	--	--	--
Spain	24 516	13 795	547	9 847	48 706	--	--	12 076	--	--	5 948
Sweden	12 100	3 200	100	700	16 100	--	--	2 800	--	--	1 500
Switzerland	9 200	2 370	185	865	12 620	75	0	75	1 112	2 432[4]	3 544
United Kingdom	--	--	--	--	109 691	--	--	5 506[2]	--	--	2 614[3]

Source: Responses to 1996 ECMT questionnaire.

See notes on next page.

Notes to Tables 12 and 13.

Austria
1. Includes agricultural/ military transport emissions.
2. Includes electric public transport.
3. Internal navigation only.
4. Domestic aviation only.
5. Public transport (excluding electric PT), tractors, road works vehicles.
6. Figures for Light Commercial Vehicles included in Private Cars category.
7. Mostly on the Danube (approx. 8 million tons/yr); tourism navigation on lakes (e.g. Bodensee) is negligible.

Belgium
1. Includes Shipping, Air and Rail emissions.

Canada
1. Includes motorcycles, alternative fuelled vehicles and diesel-powered vehicles used in public administration, industrial and farm sectors.

Czech Republic
1. Agricultural/ Military transport emissions are not included.
2. Internal navigation only.
3. Domestic aviation only.

Denmark
1. Base Year = 1988.
2. Internal navigation only.
3. Domestic aviation only.

Finland
1. Non-road mobile machinery.
2. Emissions for all ships under Finnish flag.
3. For domestic civil aviation and all civil aviation inside Finland's aviation information regions including overflights.

France
1. Data represent emissions from consumption by the transport sector of fuel sold within metropolitan France (overseas *départements* and *territoires* -- DOM-TOM -- are not included); data have been corrected for climatic variations.
2. From French and foreign ships bunkered in France, not particularly from consumption in France.

Germany
1. Figures for Light Commercial Vehicles included in Heavy Goods Vehicles and Buses category.
2. Not included in Totals figure.

Hungary
1. Agricultural/ Military transport emissions are not included.
2. Internal navigation only.
3. International flights landing in Hungary included in internal aviation.
4. Includes agricultural vehicles.

Italy
1. Diesel trains only.

Japan
1. International marine bunkers only.
2. International aviation bunkers only.

Latvia
1. Internal navigation only.
2. International aviation bunkers only; not included in transport emissions total.

Lithuania
1. Figures in ( ) represent unspecified data. They are not included in the totals.

Netherlands
1. Including electric trains.
2. Diesel trains only.
3. Off road vehicles.

New Zealand
1. All CO_2 emissions from transport are based on sectoral fuel use data. No information exists regarding how much of each fuel is used by different vehicle types, except where easily deduced, e.g. emissions from aviation sector is total of aviation fuel emissions.
2. Agricultural/Military transport emissions are not included.

Norway
1. Fishing vessels and mobile oil rigs not included.
2. Only Norwegian aircraft over Norwegian territory.
3. Figures for Light Commercial Vehicles included in Private Cars category.

Poland
1. Base year: 1995.

Portugal
1. Corresponds to the sale of diesel for road vehicles. Includes figures for "Heavy Goods Vehicles and Buses", "Motorcycles and Other" and "Light Commercial Vehicles" categories.

Romania
1. Reference year: 1993.
2. Not including military and agricultural transport emissions.
3. Figure for Light Commercial Vehicles included in Heavy Goods Vehicles and Buses category.

Russian Federation
1. Agricultural and military road vehicle emissions included if bunkered at general use petrol stations.
2. Only Russian ships and aircraft bunkered on Russian territory taken into account; military ships and aircraft not included.

Slovak Republic
1. From navigation (domestic and international) in the Slovak section of the Danube.
2. From air traffic in air corridors over Slovak Republic.
3. Includes agricultural and diesel engine machinery and military transport.
4. Includes small agricultural and other machinery with petrol engines.

Slovenia
1. "Domestic working machines".

Switzerland
1. Without correction for climatic variations.
2. Represents only emissions from diesel fuel consumed on the short Swiss stretch of the Rhine.
3. Off road transport (agriculture, military, industry).
4. Total fuel sales minus domestic consumption.

United Kingdom
1. Emissions by UNECE source category -- does not include emissions from generation of electricity to power trains.
2. Includes emissions from fishing, coastal shipping, oil exploration and production, as well as fuel oil use on offshore installations. Marine bunker emissions included only if under 12 miles from shore.
3. Includes only emissions associated with ground movement and take-off and landing cycles up to 1 km from the airport.

Table 14. Reference Case: Projected Emissions of CO_2 2000, 2010 by Sector
(with no additional measures after base year)

	TOTAL EMISSIONS (million tonnes)			TRANSPORT EMISSIONS BY SECTOR (thousand tonnes)												TRANSPORT EMISSIONS (million tonnes)			PER CENT TRANSPORT		
				Road			Rail			Shipping			Air			Others								
	1990	2000	2010	1990	2000	2010	1990	2000	2010	1990	2000	2010	1990	2000	2010	1990	2000	2010	1990	2000	2010	1990	2000	2010
Austria[1]	59.6	65	71.1	13 280	15 824	15 958	707	778	768	41[2]	45[2]	50[2]	1 002[3]	1 415[2]	1 418[3]	1 131.0[4]	1 264[4]	1 298[4]	16.2	19.3	19.5	27.0	29.7	27.4
Belgium[1]	110.0	125	125.9	21 000	29 000	33 000	1 000[2]	1 000[2]	1 000[2]	--	--	--	--	--	--	--	--	--	22.0	30.0	34.0	20.0	24.0	27.0
Czech Republic	168.6	114.9	113.9	6 840	11 305	14 072	738	380	401	54[1]	56[1]	64[1]	294[2]	1 595[2]	1 980[2]	--	--	--	7.9	13.3	16.5	4.7	11.6	14.5
Denmark[1]	53.5	54	55.3	10 577	11 223	11 509	381	378	359	221[2]	232[2]	235[2]	57[3]	66[3]	75[3]	--	--	--	11.1	11.9	12.2	21.0	22.0	22.0
Finland	65	75	--	11 200	13 500	--	240	240	--	1 600[2]	2 300[2]	2 300[2]	520	590	720	2 100[1]	2 100[1]	2 100[1]	15.6	18.7	--	24.0	25.0	--
Hungary	75.1	--	--	8 131.6	10 619.6	--	495.3	--	--	26.8[1]	--	--	532.3[2]	--	--	1 635.5	--	--	10.3	--	--	13.8	--	--
Ireland	31.9	38.5	--	4 715	5 909	--	114	92	--	117	159	--	1 111	1 425	--	0	0	--	6.1	7.5	--	19.0	19.7	--
Italy	--	463	--	--	110 400	--	--	--	--	--	1 600	--	--	2 500	--	--	5 900[1]	--	--	120.4	--	--	26.0	--
Latvia	23.6	15.7	--	3 826	3 123	--	881.4	161	--	955	50	--	168[1]	85[1]	--	--	--	--	5.6	3.3	--	24.0	21.4	--
Lithuania[1&2]	36.6	--	--	3 681	3 418	3 313	242	165	208	133	131	133	442	445	--	(1 293)	--	--	4.5 (5.8)	4.2	(6.5)	12.3	--	--
Netherlands	184.6[1]	--	--	23 800	--	35 900	100[2]	--	--	1 500	--	2 500	500	--	1 050	1 800[3]	--	2 200[3]	27.7	--	41.7	15.0	--	--
New Zealand[1]	27.9	--	--	--[2]	--[2]	--[2]	--	--	--	1 046[3]	--	--	2 156	--	--	--	--	--	11.1	--	--	40.0	--	--
Poland[1]	--	--	--	20 016	24 973	35 201	7 327	995	9 223	588	685	824	566	524	580	--	--	--	28.5	34.2	45.8	--	--	--
Romania[1]	--	--	--	6 832.2[2]	10 136.5[2]	17 390[2]	--	--	--	--	--	--	--	--	--	--	--	--	--	--	--	--	--	--
Slovak Republic[1]	53	--	--	4 500.8	4 651.8	4 922.2	376.9	187.3	180	142.6	126.2	146.3	275.9	442.7	518	--	--	--	5.3	5.4	5.7	10.0	--	--
Sweden[1]	--	--	--	--	--	--	--	--	--	--	--	--	3 544	4 456	4 903	--	--	--	--	--	--	--	--	--
Switzerland	--	--	--	--	--	--	--	--	--	--	--	--	--	--	--	--	--	--	--	--	--	--	--	--
United Kingdom	575	577	630	109 691	138 800	163 900	1 933	1 600	1 000	5 506	6 900	6 500	2 614	3 500[3]	5 100[3]	--	--	--	119.7	150.8	176.5	21.0	26.0	28.0
United States[1]	4 798.8	5 450.3	5 921.1	1 165	1 347.4	1 482.1	31.5	42.7	47.1	116.8	144.7	173.2	227.5	273.7	326.7	42.8	44.5	43.3	1 583.6[2]	1 853.1[2]	2 072.4[2]	33.0	34.0	35.0

Source: Responses to 1996 ECMT questionnaire. See notes after Table 16.

Table 15. Reference Case: Projected Emissions of CO_2 2000, 2010 by Sub-sector
(with no additional measures after base year)

(thousand tonnes)

	Private Cars			ROAD Heavy Goods Vehicles and Buses			Motorcycles and Other			Light Commercial Vehicles			TOTAL		
	1990	2000	2010	1990	2000	2010	1990	2000	2010	1990	2000	2010	1990	2000	2010
Austria[1]	8 471[5]	10 193[5]	10 329[5]	4 710	5 486	5 450	99	145	179	--[5]	--[5]	--[5]	13 280	15 824	15 958
Belgium[1]	13 000	18 000	21 000	8 000	11 000	12 000	--	--	--	--	--	--	21 000	29 000	33 000
Czech Republic	3 797	5 745	6 961	3 043[3]	5 560[3]	7 111[3]	--[3]	--[3]	--[3]	--[3]	--[3]	--[3]	6 840	11 305	14 072
Denmark[1]	6 109	6 563	6 755	2 149	2 190	2 184	--	--	--	2 319	2 496	2 570	10 577	11 223	11 509
Finland	6 850	8 500	--	3 200	3 500	--	--	--	--	1 150	1 150	--	11 200	13 500	--
Hungary	4 202.5	5 433.4	--	3 299.6	4 309.2	--	45.6[3]	59.5[3]	--	583.9	762.5	--	8 131.6	10 619.6	--
Ireland	--	--	--	--	--	--	--	--	--	--	--	--	4 715	5 909	--
Italy	--	67 000[2]	--	--	39 500	--	--	3 900	--	--[2]	--[2]	--	--	110 400	--
Latvia	1 869	1 428	--	1 903	1 600	--	54	10	--	--	--	--	3 826	3 123	--
Lithuania[1]	1 047	995	--	1 830	1 683	--	694	638	--	110	102	--	3 681	3 418	3 313
Netherlands	15 200	--	19 600	6 000	--	11 000	500	--	700	2 100	--	4 600	23 800	--	35 900
New Zealand[1]	--	--	--	--	--	--	--[2]	--[2]	--[2]	--[3]	--[3]	--[3]	--	--	--
Poland[1]	10 796	14 147	22 172	4 656	5 619	7 270	--	--	--	4 464	5 207	5 759	20 016	24 973	35 201
Romania[1]	2 979	4 703.7	7 912.2	3 780.7[3]	5 318[3]	9 284.2[3]	72.5[2]	114.8[2]	193.9[2]	--[3]	--[3]	--[3]	--	--	--
Slovak Republic[1]	1 138.4[2]	1 622.8[2]	1 812.2[2]	3 049.4[3]	3 029[3]	3 110[3]	47[2]	--[2]	--[2]	265.9[3]	--[3]	--[3]	4 500.8	4 651.8	4 922.2
Sweden[1]															
United Kingdom	--	--	--	--	--	--	--	--	--	--	--	--	109 691	138 800	163 900
United States[1]	897.9[3]	1 030.9[3]	1 114.8[3]	267.1	316.5	367.3	--	--	--	--[3]	--[3]	--[3]	1 165	1 347.4	1 482.1

Source: Responses to 1996 ECMT questionnaire. See notes after Table 16.

Table 16. Reference Case: Projected Emissions of CO_2 2000, 2010 by Sub-sector
(with no additional measures after base year)

(thousand tonnes)

	SHIPPING									AVIATION								
	Internal Navigation			International Marine Bunkers			TOTAL			Domestic Aviation			International Aviation Bunkers			TOTAL		
	1990	2000	2010	1990	2000	2010	1990	2000	2010	1990	2000	2010	1990	2000	2010	1990	2000	2010
Austria[1]	41	45	50	--	--	--	41	45	50	1 002	1 415	1 418	--	--	--	1 002	1 415	1 418
Belgium[1]							1 000 [2]	1 000 [2]	1 000 [2]							1 000 [2]	1 000 [2]	1 000 [2]
Czech Republic	54	56	64	--	--	--	54	56	64	294	1 595	1 980	--	--	--	294	1 595	1 980
Denmark[1]	221	232	235	--	--	--	221	232	235	57	66	75	--	--	--	57	66	75
Finland	--	--	--	1 600	2 300	2 300	1 600	2 300	2 300	220	260	310	--	--	--	520	590	720
Hungary	26.8	--	--	--	--	--	26.8	--	--							532.3	--	--
Ireland	56	49	--	61	110	--	117	159	--	--	--	--	1 111	1 425	--	1 111	1 425	--
Italy	--	600	--	--	1 000	--	--	1 600	--	--	2 500	--	--	--	--	--	2 500	--
Latvia	955	50	--	--	--	--	955	50	--	--	--	--	168	85	--	168	85	--
Lithuania[1]	13	12	--	120	119	133	133	131	133	8	20	--	434	425	--	442	445	--
Netherlds	--	--	--	1 500	--	2 500	1 500	--	2 500	--	--	--	500	--	--	500	--	1 050
New Zeal.[1]	--	--	--	1 046	--	--	1 046	--	--	789	--	--	1 367	--	--	2 156	--	--
Poland[1]	--	--	--	588	--	--	588	685	824	--	--	--	--	--	--	566	524	580
Slovak Republic[1]	142.6	--	--	--	--	--	142.6 [4]	126.2 [4]	146.3 [4]	--	--	--	--	--	--	275.9 [5]	442.7 [5]	518 [5]
Sweden[1]																		
Switzerland	--	--	--	--	--	--	--	--	--	1 112	1 243	1 367	2 432 [1]	3 213 [1]	3 534 [1]	1 112	1 243	1 367
United Kingdom	--	--	--	5 506 [2]	6 900 [2]	6 500 [2]	5 506 [2]	6 900 [2]	6 500 [2]	--	--	--	--	--	--	2 614	3 500 [3]	5 100 [3]
U.S.[1]	39.1	45.8	49.4	77.8	98.9	123.8	116.8	144.7	173.2	--	--	--	--	--	--	227.5	273.7	326.7

Source: Responses to 1996 ECMT questionnaire. See notes on next page.

Notes to Tables 14, 15 and 16.

Austria
1. Modelling exercise undertaken in 1995.
2. Internal navigation only.
3. Domestic aviation only.
4. Public transport (excluding electric PT), tractors, road works vehicles.
5. Figures for agricultural and military transport emissions and Light Commercial Vehicles included in Private Car category.

Belgium
1. Projected emissions are for 2000 and 2005.
2. Includes Rail, Shipping, and Air emissions.

Czech Republic
1. Internal navigation only.
2. Domestic aviation only.
3. Figures for "Heavy Goods Vehicles and Buses", "Motorcycles and Other", and "Light Commercial Vehicles" grouped together. Agricultural and military transport emissions are not included.

Denmark
1. Base year: 2000; projected years: 2005 and 2010.
2. Internal navigation only.
3. Domestic aviation only.

Finland
1. Non-road mobile machinery.

Hungary
1. Internal navigation only.
2. Domestic aviation only.
3. Figures for agricultural and military transport not included.

Italy
1. Includes public sector and utility vehicles.
2. Figures for Light Commercial Vehicles included in Private Cars category.

Latvia
1. International aviation bunkers.

Lithuania
1. Projected emissions are for 2000 and 2005.
2. Figures in ( ) represent unspecified data. They are not included in totals.

Netherlands
1. Including electric trains.
2. Diesel trains only.
3. Off-road vehicles.

New Zealand
1. All CO_2 emissions from transport are based on sectoral fuel use data. No information exists regarding how much of each fuel is used by different vehicle types, except where easily deduced, e.g. emissions from aviation sector is total of aviation fuel emissions.
2. Agricultural and military vehicles not included.
3. International Marine Bunker emissions only.

Poland
1. Base year: 1995.

Romania
1. Base year: 1993.
2. Not including military and agricultural transport emissions.
3. Figures for Light Commercial Vehicles included in Heavy Goods Vehicles and Buses category.

Slovak Republic
1. Projected emissions are for 2000 and 2005.
2. Figures for Motorcycles and Other included in Private Car category.
3. Figures for Light Commercial Vehicles included in Heavy Goods Vehicles and Buses category.
4. From navigation (domestic and international) in the Slovak section of the Danube.
5. From air traffic in air corridors over Slovak Republic.

Sweden
1. Questionnaire notes that data approximates that found in Table 8: Status Quo Case.

Switzerland
1. Total fuel sales minus domestic consumption.

United Kingdom
1. Emissions by UNECE source category -- does not include emissions from generation of electricity to power trains.
2. Includes emissions from fishing, coastal shipping, oil exploration and production, as well as fuel oil use on offshore installations. Marine bunker emissions included only if under 12 miles from shore.
3. Includes only emissions associated with ground movement and take-off and landing cycles up to 1 km from the airport.

United States
1. All emissions figures expressed in million tonnes.
2. Total includes military fuel use.
3. Figures for Light Commercial Vehicles included in Private Cars category.

Table 17. Status Quo Case: Projected Emissions of CO_2 2000, 2010 by Sector
(with only measures implemented since base year)

	TOTAL EMISSIONS (million tonnes)			TRANSPORT EMISSIONS BY SECTOR (thousand tonnes)												TRANSPORT EMISSIONS (million tonnes)			PER CENT TRANSPORT		
				Road			Rail			Shipping			Air			Others								
	1990	2000	2010	1990	2000	2010	1990	2000	2010	1990	2000	2010	1990	2000	2010	1990	2000	2010	1990	2000	2010	1990	2000	2010
Canada	451.9	484.7	536.9	117 800	131 300	145 500	6 300	6 400	7 000	5 700	5 700	6 300	10300	11 600	13 000	--	--	--	140.1	155.1	171.8	31.0	32.0	32.0
Czech Republic[1]	168.6	115.4	113.9	6 840	9 735	11 645	738	340	366	54	50[2]	59[2]	294[3]	1 071[3]	1 376[3]	--	--	--	7.9	11.2	13.4	4.7	9.7	11.8
Finland	65.0	75.4	71.0	11 200	11 300	11 600	240	270	270	1 600	2 300	2 300	520	590	720	2 100	2 100	2 100	15.6	16.6	17.0	24.0	22.0	24.0
Lithuania[1&2]	--	--	--	3 681	3 550	5 079	242	230	242	133	129	176	442	445	--	(1 293)	--	--	4.5 (5.8)	4.3	(6.5)	12.3	--	--
Netherlands[1]	184.6	--	--	23 800	--	30 600	100	--	--	1 500	--	2 100	500	--	1 050	1 800[2]	--	2 200[2]	27.7	--	36.0	15.0	--	--
New Zealand[1]	27.9	36.3	42.9	--	11 715	14 243	--	--	--	--	1 634	1 798	2 156	2 263	2 820	--	--	--	11.1	15.6	18.8	40.0	43.0	44.0
Norway[1]	32.4	--	--	8 000	7 995	9 619	100	110	110	2 000	1 253	1 371	1 300[2]	1 400[2]	1 814[2]	--	--	--	11.4	--	--	35.2	--	--
Romania[1]	--	--	--	6 832	10 137[2]	17 390[2]	--	--	--	--	--	--	--	--	--	--	--	--	--	--	--	--	--	--
Slovak Republic[1]	53	46	47.2	4 501	4 326	4 360	377	187	180	142.6[2]	135[2]	160[2]	276	402	500	--	--	--	5.3	5.0	5.2	10.0	11.0	11.0
Sweden	61.3	63.8	--	16 100	17 400	18 900	100	100	200	2 800	2 500	2 500	1 500	1 800	1 800	--	--	--	20.5	21.8	23.4	33.4	34.0	--
Switzerland	44.2	44.9[1]	46.6[1]	12 620	13 860	15 385	30	30	25	75[2]	70[2]	65[2]	1 112	1 243	1 367	975[3]	940[3]	1 050[3]	14.7	16.3	18.3	33.4	36.4	39.3
United States[1]	4 798.8	5 216.5	5 618.3	1 165	1 322	1 436	32	43	47	116.8	144	172.5	228	274	327	42.8	43.5	40.1	1 583.6[2]	1 825.8[2]	2 022.6[2]	33.0	35.0	36.0

Source: Responses to 1996 ECMT questionnaire.

See notes after Table 19.

Table 18. **Status Quo Case: Projected Emissions of CO_2 2000, 2010 by Sub-sector**
(with only measures implemented since base year)

(thousand tonnes)

	ROAD												TOTAL		
	Private Cars			Heavy Goods Vehicles and Buses			Motorcycles and Other			Light Commercial Vehicles					
	1990	2000	2010	1990	2000	2010	1990	2000	2010	1990	2000	2010	1990	2000	2010
Canada	52 200	54 000	54 500	23 200	32 000	36 400	20 100	18 800 [1]	20 500 [1]	22 300	26 500	34 100	117 800	131 300	145 500
Czech Republic	3 797	4 776	5 554	3 043 [4]	4 959 [4]	6 091 [4]	[4]	[4]	[4]	[4]	[4]	[4]	6 840	9 735	11 645
Finland	6 850	6 700	6 400	3 200	3 300	3 650	--	--	--	1 150	1 300	1 550	11 200	11 300	11 600
Lithuania[1]	1 047	995	--	1 830	1 775	--	694	673	--	110	107	--	3 681	3 550	5 079
Netherlands[1]	15 200	--	16 200	6 000	--	9 800	500	--	700	2 100	--	3 900	23 800	--	30 600
New Zealand[1]	--	--	--	--	--	--	--[2]	--[2]	--[2]	[3]	[3]	[3]	--	11 715	14 243
Norway[1]	5 800 [3]	--	--	2 100	--	--	--	--	--	[3]	[3]	[3]	8 000	7 995	9 619
Romania[1]	2 979	4 703.7	7 912.2	3 780.7 [3]	5 318 [3]	9 284.2 [3]	72.5 [4]	114.8 [4]	193.9 [4]	[3]	[3]	[3]	6 832	10 137	17 390
Slovak Republic[1]	1 138.4 [3]	1 509.2 [3]	1 550 [3]	3 049.4 [4]	2 817 [4]	2 810 [4]	--	--[3]	--[3]	--	--[4]	--[4]	4 501	4 326	4 360
Sweden	12 100	13 000	14 200	3 200	3 500	3 800	100	100	100	700	800	800	16 100	17 400	18 900
Switzerland	9 200	9 790	10 000	2 370	2 820	3 855	185	185	190	865	1 065	1 340	12 620	13 860	15 385
United States[1]	897.9 [3]	1 009.1 [3]	1 076.1 [3]	267.1	312.9	360.1	--	--	--	--[3]	--[3]	--[3]	1 165	1 322	1 436

Source: Responses to 1996 ECMT questionnaire.

See notes after Table 19.

Table 19. **Status Quo Case: Projected Emissions of CO_2 2000, 2010 by Sector**
(with only measures implemented since base year)

(thousand tonnes)

	SHIPPING												AVIATION									
	Internal Navigation			International Marine Bunkers			TOTAL			Domestic Aviation			International Aviation Bunkers			TOTAL		
	1990	2000	2010	1990	2000	2010	1990	2000	2010	1990	2000	2010	1990	2000	2010	1990	2000	2010
Canada	5 700	5 700	6 300	--	--	--	5 700	5 700	6 300	10 300	11 600	13 000	--	--	--	10 300	11 600	13 000
Czech Republic	54	50	59	--	--	--	54	50	59	294	1 071	1 376	--	--	--	294	1 071	1 376
Finland	--	--	--	1 600	2 300	2 300	1 600	2 300	2 300	220	260	310	520	590	720	520	590	720
Lithuania[1]	13	12	--	120	117	--	133	129	176	8	20	--	434	425	--	442	445	--
Netherlands[1]	--	--	--	1 500	--	--	1 500	--	2 100	--	--	--	--	--	--	500	--	1 050
New Zealand[1]	--	693	758	1 046	941	1 040	--	1 634	1 798	789	900	1 056	1 367	1 363	1 764	2 156	2 263	2 820
Norway[1]	2 000[4]	1 253	1 371	--	--	--	2 000[4]	1 253	1 371	--	--	--	--	--	--	1 300[2]	1 400	1 814
Slovak Republic[1]	142.6[5]	--	--	--	--	--	142.6	135	160	--	--	--	--	--	--	275.9[6]	401.5	500
Sweden	--	--	--	2 800	--	--	2 800	2 500	2 500	--	--	--	--	--	--	1 500	1 800	1 800
Switzerland	75	70	65	0	0	0	75[2]	70[2]	65[2]	1 112	1 243	1 367	2 432[4]	3 213[4]	3 534[4]	1 112	1 243	1 367
United States[1]	39.1	45.1	48.7	77.8	98.9	123.8	116.8	144	172.5	--	--	--	--	--	--	227.5	273.7	326.7

Source: Responses to 1996 ECMT questionnaire.

See notes on next page.

146

Notes to Tables 17, 18 and 19.

Canada
1. Includes motorcycles, alternative fuelled vehicles and diesel-powered vehicles used in public administration, industrial and farm sectors.

Czech Republic
1. Emissions are for 1990, 2000 and 2005.
2. Internal navigation only.
3. Domestic aviation only.
4. Includes figures for "Heavy Goods Vehicles and Buses", "Motorcycles and Other" and "Light Commercial Vehicles" categories. Agricultural and military transport emissions are not included.

Lithuania
1. Emissions are for 1990, 2000 and 2005.
2. Figures in ( ) represent unspecified data, they are not included in totals.

Netherlands
1. Only measures implemented during the period 1990-1995.
2. Off road vehicles.

New Zealand
1. All CO_2 emissions from transport are based on sectoral fuel use data. No information exists regarding how much of each fuel is used by different vehicle types, except where easily deduced, e.g. emissions from aviation sector is total of aviation fuel emissions.
2. Agricultural and military transport emissions are not included.

Norway
1. Projected figures are not comparable with base year data: base year = CO_2 by source; projected years = by CO_2 by sector.
2. Only Norwegian aircraft over Norwegian territory.
3. Figures for Light Commercial Vehicles included in Private Cars category.
4. Not including fishing vessels and mobile oil rigs.

Romania
1. Reference year: 1993.
2. Not including military and agricultural transport emissions.
3. Figures for Light Commercial Vehicles included in Heavy Goods Vehicles and Buses and category.
4. Not including military and agricultural transport emissions.

Slovak Republic
1. Emission projections are for 2000 and 2005.
2. Internal navigation only.
3. Figures for Motorcycles and Other (including small agricultural and other machinery with petrol engines) included in Private Car category.
4. Figures for Light Commercial Vehicles, agricultural and diesel engine machinery and military transport included in Heavy Goods Vehicles and Buses category.
5. From navigation (domestic and international) in the Slovak section of the Danube.
6. From traffic in air corridors over the Slovak Republic.

Switzerland
1. Without correction for climate.
2. Represents only emissions from diesel fuel consumed on the short Swiss stretch of the Rhine.
3. Off road transport (agriculture, military and industry).
4. Total fuel sales minus domestic consumption.

United States
1. All emissions figures expressed in million tonnes.
2. Total includes military fuel use.
3. Figures for Light Commercial Vehicles included in Private Car category.

Table 20. **Future Measures Case: Projected Emissions of CO_2, 2000, 2010 by Sector**
(with measures already implemented or planned for implementation)

	TOTAL EMISSIONS (million tonnes)			TRANSPORT EMISSIONS BY SECTOR (thousand tonnes)												TRANSPORT EMISSIONS (million tonnes)			PER CENT TRANSPORT		
				Road			Rail			Shipping			Air			Others								
	1990	2000	2010	1990	2000	2010	1990	2000	2010	1990	2000	2010	1990	2000	2010	1990	2000	2010	1990	2000	2010	1990	2000	2010
Austria[1]	59.6	65	71.0	13 280	15 824	14 503	707	778	768	41[2]	45[2]	50[2]	1 002[3]	1 415[3]	1 418[3]	1 131	1 264	1 298	16.2	19.3	18.0	27.1	29.7	25.4
Belgium[1]	110	112.2	119	21 000	26 500	30 000	1 000[2]	1 000[2]	1 000[2]	[2]	[2]	[2]	[2]	[2]	[2]	--	--	--	22.0	27.5	30.0	20.0	24.5	25.2
Czech Republic	168.6	115.2	114.6	6 840[1]	8 959[1]	9 281[1]	738	340	280	54[2]	50[2]	57[2]	294[3]	901[3]	925[3]	--	--	--	7.9	10.2	10.5	4.7	8.9	9.2
Denmark[1]	60.2	48.7	--	9 239	9 550	--	428	380	--	445	230	--	67	65	--	--	--	--	10.2	10.2	--	17.0	21.0	--
Finland	65	--	--	11 200	--	--	240	--	--	1 600	--	--	520	590	720	2 100	--	--	15.6	--	--	24.0	--	--
Lithuania[1&2]	36.6	--	--	3 681	3 550	6 699	242	230	290	133	129	230	442	445	--	(1 293)	--	--	4.5 (5.8)	4 354	(6 544)	12.3	--	--
Netherlands	184.6	--	--	23 800	--	25 700	100	--	--	1 500	--	2 100	500	--	1 050	1 800[1]	--	1 500[1]	27.7	--	30.4	15.0	--	--
New Zealand[1]	27.9	--	--	--[2]	--[2]	--[2]	--	--	--	1 046[3]	--	--	3 523	--	--	--	--	--	11.1	--	--	40.0	--	--
Poland[1]	--	--	--	20 016	22 956	29 969	7 327	7 995	9 223	588	685	824	567	523	580	--	--	--	28.5	32.2	40.6			
Romania[1]	--	--	--	6 832.2	9 157.6	14 933.6	--	--	--	--	--	--	--	--	--	--	--	--	--	--	--	--	--	--
Slovak Republic[1]	53	--	--	4 500.8	4 150	4 090	376.9	180	170	142.6[2]	146[2]	170[2]	275.9[3]	390[3]	480[3]	--	--	--	5.3	4.9	4.9	10.0	--	--
Sweden	61.3	--	--	16 100	16 500	14 500	100	100	200	2 800	2 500	2 500	1 500	1 800	1 800	--	--	--	20.5	20.9	19.0	33.4	--	--
Switzerland	--	--	--	--	--	--	--	--	--	--	--	--	3 544	4 365	4 600	--	--	--	--	--	--	--	--	--
United Kingdom	575	545	595	109 691	125 400	141 900	1 933[1]	1 600[1]	1 000[1]	5 506[2]	6 900[2]	6 500[2]	2 614[3]	3 500[3]	5 100[3]	--	--	--	119.7	137.3	154.4	21.0	25.0	27.0

Source: Responses to 1996 ECMT questionnaire.

See notes after Table 22.

Table 21. **Future Measures Case: Projected Emissions of CO_2 2000, 2010 by Sub-sector**
(with measures already implemented or planned for implementation)

(thousand tonnes)

	ROAD												TOTAL		
	Private Cars			Heavy Goods Vehicles and Buses			Motorcycles and Other			Light Commercial Vehicles					
	1990	2000	2010	1990	2000	2010	1990	2000	2010	1990	2000	2010	1990	2000	2010
Austria	8 471 [4]	10 193 [4]	9 234 [4]	4 710	5 486	5 090				[4]	[4]	[4]	13 280	15 824	14 503
Belgium[1]	13 000	16 500	19 000	8 000	10 000	11 000	--	--	--	--	--	--	21 000	26 500	30 000
Czech Republic[1]	3 797	4 437	4 378	3 043 [4]	4 522 [4]	4 903 [4]	[4]	[4]	[4]	[4]	[4]	[4]	6 840	8 959	9 281
Denmark[1]	5 013	5 300	--	2 089	2 050	--	--	--	--	2 137	2 200	--	9 239	9 550	--
Lithuania[1]	1 047	995	--	1 830	1 775	--	694	675	--	110	107	--	3 681	3 550	6 699
Netherlands	15 200	--	14 400	6 000	--	7 800	500	--	700	2 100	--	2 800	23 800	--	25 700
New Zealand[2]	--	--	--	--	--	--	--	--	--	--	--	--	--	--	--
Poland[1]	10 796 [2]	12 420 [2]	18 233 [2]	4 756	5 460	6 553	[2]	[2]	[2]	4 464 [3]	5 076 [3]	5 183 [3]	20 016	22 956	29 969
Romania[1]	2 979	3 757.9	6 111.3	3 780.7 [2]	5 308 [2]	8 675.6 [2]	72.5 [3]	91.7 [3]	146.7 [3]	[2]	[2]	[2]	6 832.2	9 157.6	14 933.6
Slovak Republic[1]	1 138.4 [4]	1 450 [4]	1 390 [4]	3 049.4	2 700 [5]	2 700 [5]	--	[4]	[4]	--	[6]	[6]	4 500.8	4 150	4 090
Sweden	12 100	12 400	10 900	3 200	3 300	2 900	100	100	100	700	700	600	16 100	16 500	14 500
United Kingdom	--	--	--	--	--	--	--	--	--	--	--	--	109 691	125 400	141 900

Source: Responses to 1996 ECMT questionnaire.

See notes after Table 22.

Table 22. **Future Measures Case: Projected Emissions of CO$_2$ 2000, 2010 by Sub-sector**
(with measures already implemented or planned for implementation)

(thousand tonnes)

| | SHIPPING ||||||||| AVIATION |||||||||
| | Internal Navigation ||| International Marine Bunkers ||| TOTAL ||| Domestic Aviation ||| International Aviation Bunkers ||| TOTAL |||
	1990	2000	2010	1990	2000	2010	1990	2000	2010	1990	2000	2010	1990	2000	2010	1990	2000	2010
Austria	41	45	50	--	--	--	41	45	50	1 002	1 415	1 418	--	--	--	1 002	1 415	1 418
Belgium[1]																1 000 [2]	1 000 [2]	1 000 [2]
Czech Republic	54	50	57	--	--	--	54	50	57	294	901	925	--	--	--	294	901	925
Denmark[1]	445	230	--	--	--	--	445	230	--	67	65	--	--	--	--	67	65	--
Finland	--	--	--	--	--	--	1 600	--	--	220	260	310	--	--	--	520	590	720
Lithuania[1]	13	12	--	120	117	--	133	129	230	8	20	--	434	425	--	442	445	--
Netherlands	--	--	--	--	--	--	1 500	--	2 100	--	--	--	--	--	--	500	--	1 050
New Zealand	--	--	--	1 046	--	--	1 046	--	--	1 367	--	--	2 156	--	--	3 523	--	--
Slovak Republic[1]	142.6	150	170	--	--	--	142.6 [2]	146 [2]	170 [2]	--	--	--	--	--	--	275.9 [3]	390 [3]	480 [3]
Sweden	--	--	--	--	--	--	2 800	2 500	2 500	--	--	--	--	--	--	1 500	1 800	1 800
Switzerland	--	--	--	--	--	--	--	--	--	1 112	1 215	1 300	2 432 [1]	3 150 [1]	3 300 [1]	3 544	4 365	4 600
United Kingdom	--	--	--	--	--	--	5 506 [2]	6 900 [2]	6 500 [2]	--	--	--	--	--	--	2 614 [3]	3 500 [3]	5 100 [3]

Source: Responses to 1996 ECMT questionnaire.

See notes on next page.

Notes to Tables 20, 21 and 22.

Austria
1. Assumes that in 2003, road pricing will be introduced on motorways and by 2010 on all highways. A vignette will be introduced for use of all motorways 01/01/97.
2. Internal navigation only.
3. Domestic aviation only.
4. Figures for Light Commercial Vehicles included in Private Cars category. Figures also include agricultural and military transport emissions.

Belgium
1. Projected emissions are for 2000 and 2005.
2. Includes Shipping, Air and Rail emissions.

Czech Republic
1. Agricultural and military transport emissions are not included.
2. Internal navigation only.
3. Domestic aviation only.
4. Includes figures for the "Heavy Goods Vehicles and Buses", "Motorcycles and Other" and "Light Commercial Vehicles" categories.

Denmark
1. Emissions are for 2000, 2005 and 2010.

Lithuania
1. Projected emissions are for 2000 and 2005.
2. Figures in ( ) represent unspecified data. They are not included in totals.

Netherlands
1. Off road vehicles.

New Zealand
1. All CO_2 emissions from transport are based on sectoral fuel use data. No information exists regarding how much of each fuel is used by different vehicle types, except where easily deduced, e.g. emissions from aviation sector is total of aviation fuel emissions.
2. Agricultural and military transport emissions are not included.
3. International Marine Bunker emissions only.

Poland
1. Base year: 1995.
2. Figures for motorcycles included in "Private Cars" category.
3. Vehicles upto 3.5 tonnes.

Romania
1. Base year: 1993.
2. Figures for Light Commercial Vehicles included in Heavy Goods Vehicles and Buses category.
3. Not including military and agricultural transport emissions.

Slovak Republic
1. Projected emissions are for 2000 and 2005.
2. From navigation (domestic and international) in the Slovak section of the Danube.
3. From traffic in air corridors over the Slovak Republic.
4. Figures for Motorcycles and Other (including small agricultural and other machinery with petrol engines) included in the Private Cars category.
5. Includes Light Commercial Vehicles, agricultural and diesel machinery and military transport.
6. Figure included in Heavy Goods Vehicles and Buses category.

Switzerland
1. Total fuel sales minus domestic consumption.

United Kingdom
1. Emissions by UNECE source category -- does not include emissions from generation of electricity to power trains.
2. Includes emissions from fishing, coastal shipping, oil exploration and production, as well as fuel oil use on offshore installations. Marine bunker emissions included only if under 12 miles from shore.
3. Includes only emissions associated with ground movement and take-off and landing cycles up to 1 km from the airport.

(thousand tonnes)

Table 23. **Emissions of CO_2 2000, 2010 from Road Transport by Fuel**
(with only measures Implemented or Agreed)

	TOTAL 2000									TOTAL 2010[1]							
	Petrol			Diesel[1]			Other			Petrol			Diesel[1]			Other		
	Reference	Status Quo	Future	Reference	Status Quo	Future	Reference	Status Quo	Future	Reference	Status Quo	Future	Reference	Status Quo	Future	Reference	Status Quo	Future
Austria	7 006	7 006	--	HGV:10 083	HGV:10 083	--	217[5]	217[5]	--	--	--	5 177	--	--	9 611	--	--	217[5]
Belgium	9 000	--	--	HGV: 8 500 LV: 3 500	--	--	--	--	--	9 000	--	--	HGV: 8 500 LV: 3 500	--	--	--	--	--
Canada	--	85 000	--	--	HGV:24 300 LV: 5 700	--	--	16 200[12]	--	--	93 700	--	--	HGV:27 200 LV: 6 500	--	--	18 000[12]	--
Czech Republic	4 981	4 141	3 831	HGV: 5 184	4 639	4 221	--	--	242[2]	6 314	4 907	2 642	HGV: 7 058	6 135	5 071	--	--	1 089[2]
Denmark	5 928	--	--	HGV: 2 149 LV: 2 499	--	--	--	--	--	6 354	--	5 150	HGV: 2 190 LV: 2 679	--	2 050 2 350	--	--	--
Lithuania[7]	1 246	1 717	2 078	847	1 166	1 411	68[6]	65[6]	78[6]	1 770	2 494	3 288	1 202	1 693	2 233	66[6]	94[6]	124[6]
Latvia	2 667	--	--	HGV: 565 LV: 42	--	--	60	--	--	--	--	--	--	--	--	--	--	--
Netherlands	--	--	--	--	--	--	--	--	--	16 500	14 600	12 000	HGV 11 500 LV 10 100[10]	HGV: 9 800 LV: 8 700[10]	HGV 8 400 LV 7 400[10]	3 600[11]	3 000[11]	2 600[11]
New Zealand	--	7 506	--	--	3 889[3]	--	--	320[4]	--	--	9 213	--	--	4 803[3]	--	--	226[4]	--
Romania	5 107.2	5 107.2	4 078.9	HGV 5 029.3	HGV 5 029.3	HGV 5 078.7				8 624.4	8 624.4	6 527.1	HGV 8 765.9	HGV 8 765.9	HGV 8 406.5			
Slovak Republic	1 600	1 480	1 400	3 042	2 825[3]	2 711[3]	10[4]	21[4]	39[4]	1 780	1 500	1 290	3 116	2 703	2 661	26[4]	67[4]	139[4]
Sweden	--	13 200	--	--	4 200[3]	--	--	--	--	--	14 600	--	--	4 300[3]	--	--	--	--
Switzerland	--	10 000	--	--	HGV:28 205 LV: 10 402	--	--	0	--	--	10 235	--	--	HGV:38500 LV: 1 300	--	--	0	--
United Kingdom	85 226	--	76 040	53 574	--	49 360	--	--	--	87 016	--	71 750	76 885	--	70 150	--	--	--
United States	1 083.9	1 061.3	--	HGV: 264.3 LV: 10.9 Rail: 69.5	HGV: 261.4 LV: 10.1 Rail: 69.5	--	273.7[8] 150.8[9]	273.7[8] 149.7[9]	--	1 131.8	1 093.7	--	HGV: 321.5 LV: 13.0 Rail: 76.8	HGV: 315.0 LV: 12.3 Rail: 76.0	--	326.7[8] 202.6[9]	326.7[8] 198.8[9]	--

Source: Responses to 1996 ECMT questionnaire.

Notes to Table 23: Emissions of CO_2 2000, 2010 from Road Transport by Fuel (with only measures Implemented or Agreed)

1. HGV = Heavy Goods Vehicles.
2. LV = Light Vehicles.
3. Other = Methanol for private cars.
4. Diesel = both HGV and LV.
5. Other = Natural Gas and Liquid Petroleum Gas.
6. Other = Fuel for agricultural vehicles.
7. Other = Liquid Petroleum Gas.
8. Emissions projections are for 2005 instead of 2010.
9. Jet fuel.
10. Natural gas, Liquid Petroleum Gas, Residual oil.
11. Light Vehicles include figures for off road vehicles.
12. Other = LPG and kerosene.
13. Includes motor cycles, alternative fuels, diesel-powered vehicles used in public administration, industrial and farm sectors.

Table 24. Annual CO_2 Emissions Data from Road Transport

(thousand tonnes)

	ROAD																		
	Private Cars						Heavy Goods Vehicles and Buses						Motorcycles and Other					
	1990	1991	1992	1993	1994	1995	1990	1991	1992	1993	1994	1995	1990	1991	1992	1993	1994	1995
Austria	8 471	9 400	9 183	9 117	9 113	9 615	4 710	5 247	5 461	5 251	5 705	5 545	99	106	105	107	112	124
Belgium	13 000	12 500	14 000	14 500	15 000	15 500	8 000	8 000	9 000	9 500	10 000	10 500	--	--	--	--	--	--
Canada	52 200	51 000	52 500	53 700	54 200	54 200	23 200	22 300	22 600	24 200	27 300	29 800	20 100	18 700	17 900	18 400	19 600	20 600
Czech Republic	3 797	3 508	3 810	4 002	4 188	4 422	3 043	3 123	3 487	4 028	4 471	4 501						
Denmark[1]	5 014	5 426	5 508	5 548	5 669	5 631	2 089	2 122	2 129	2 042	2 021	2 036	--	--	--	--	--	--
Finland	6 850	6 800	6 800	6 250	6 300	6 250	3 200	2 950	2 900	2 900	3 150	3 100	--	--	--	--	--	--
France[1]	64 600	65 500	67 200	68 000	69 100	69 400	45 200[2]	47 000[2]	48 000[2]	48 100[2]	48 400[2]	48 900[2]	1 700	1 700	1 700	1 700	1 700	1 700
Germany	110 800	--	--	--	--	--	38 500	--	--	--	--	--	900	--	--	--	--	--
Hungary	4 202.5	--	--	3 843.7	3 658.3	--	3 299.6	--	--	2 735.8	2 601.9	--	45.6	--	--	--	--	--
Italy	53 700	--	--	61 700	61 100	--	35 000	--	--	34 900	34 800	--	2 600	--	--	3 100	3 300	--
Latvia	1 869	973	934	756	726	952	1 903.1	1 177	1 003	1 039	1 060	936	54	30	29	2	2	2
Lithuania	1 143	--	--	534	--	--	2 448	--	--	1 982	--	--	--	--	--	--	--	--
Netherlands	15 200	15 200	15 700	15 900	16 300	17 200	6 000	6 700	7 000	6 700	6 700	7 200	500	200	200	300	300	300
New Zealand	--	--	--	--	--	--												
Norway		5 900	5 700	5 900	6 000	--		2 100	2 200	2 600	2 400	--	--	100	100	100	100	--
Poland	5 929[1]	--	--	--	--	10 796[1]	5 310	--	--	--	--	4 756	--	--	--	--	--	--
Romania	--	--	--	2 979	3 495.9	3 689.2	--	--	--	3 780.7	3 949.4	4 096.1	--	--	--	72.5	85.2	89.9
Russian Federation	12 000	14 700	15 100	16 700	--	--	111 700	114 900	97 700	83 300	--	--	--	--	--	--	--	--
Slovak Republic	1 138.4	1 133.7	1 282.1	1 438.7	1 370.2	--	3 049.4	2 837.2	1 938.5	1 820	2 044	--	47.1	48.7	35.5	49.6	49.6	--
Sweden	12 100	--	--	11 900	--	--	3 200	--	--	3 200	--	--	100	--	--	100	--	--
Switzerland	9 200	9 240	9 290	9 340	9 480	9 520	2 370	2 380	2 390	2 390	2 400	2 440	185	185	180	180	180	180
United States[1&2]							260[4]	253[4]	259[4]	274[4]	292[4]	294[4]	--	--	--	--	--	--

Source: Responses to 1996 ECMT questionnaire.   See notes after Table 29.

155

Table 24 cont.d. **Annual CO_2 Emissions Data from Road Transport**

(thousand tonnes)

	ROAD						TOTAL ROAD TRANSPORT					
	Light Commercial Vehicles											
	1990	1991	1992	1993	1994	1995	1990	1991	1992	1993	1994	1995
Austria	--	--	--	--	--	--	13 280	14 753	14 749	14 475	14 930	15 284
Belgium[1]	8 000[1]	8 000[1]	9 000[1]	9 500[1]	10 000[1]	10 500[1]	21 000	20 500	23 000	24 000	25 000	26 000
Canada	22 300	21 300	21 400	22 200	24 000	23 400	117 800	113 300	114 400	118 500	125 100	128 000
Czech Republic	3 043[1]	3 123[1]	3 487[1]	4 028[1]	4 471[1]	4 501[1]	6 840	6 631	7 297	8 030	8 659	8 923
Denmark[1]	2 137	2 139	2 137	2 161	2 260	2 297	9 240	9 688	9 774	9 751	9 951	9 964
Finland	1 150	1 100	1 150	1 050	1 100	1 050	11 200	10 850	10 850	10 200	10 550	10 400
France[1]	45 200[2]	47 000[2]	48 000[2]	48 100[2]	48 400[2]	48 900[2]	111 500	114 200	116 900	117 800	119 200	120 000
Germany	--	--	--	--	--	--	150 000	--	--	--	--	--
Hungary	583.9	--	--	--	--	--	8 131.6	7 145.4	6 902.8	6 579.5	6 260.2	--
Ireland	--	--	--	--	--	--	4 715	4 902	4 973	5 335	5 495	5 921[1]
Italy	53 700[1]	--	--	61 700[1]	61 100[1]	--						
Latvia	--	--	--	--	--	--	3 826.1	2 180	1 966	1 797	1 788	1 890
Lithuania	--	--	--	--	--	--	3 591	--	--	2 516	--	--
Netherlands	2 100	2 200	2 300	2 600	2 700	2 700	23 800	24 400	25 200	25 500	26 000	27 400
New Zealand	--	--[1]	--[1]	--[1]	--[1&2]	--	--	--	--	--	--	--
Norway	--	--	--	--	--	--	--	8 100	8 100	8 600	8 500[2]	--
Poland	4 177[2]	--	--	--	--	4 464[2]	15 416	--	--	--	--	20 016
Romania	--	--	--	3 780.7[1]	3 949.4[1]	4 096.1[1]	--	--	--	6 832.2	7 530.5	7 875.2
Russian Federation	23 200	24 200	20 400	17 100	--	--	146 900	153 800	133 200	117 200	102 000	92 900
Slovak Republic	265.9		1 938.5	1 820	2 044	--	4 500.8	4 019.6	3 495.1	3 436	3 651.3	--
Slovenia	--	--	--	--	--	--	2 947	2 968	3 122	3 699	3 930	4 327
Sweden	700	--	--	700	--	--	16 100	--	--	15 900	--	--
Switzerland	865	870	875	880	890	920	12 620	12 675	12 735	12 790	12 950	13 060
United Kingdom	--	--	--	--	--	--	109 691	108 911	110 359	111 533	112 193	--
United States[1&2]	898[3]	892[3]	908[3]	935[3]	940[3]	964[3]	1 158	1 145	1 167	1 209	1 232	1 258
European Union[1]							639 595.7	653 739.7	673 522.3	683 521.4	688 810.4	--

Source: Responses to 1996 ECMT questionnaire.

See notes after Table 29.

Table 25. Annual CO$_2$ Emissions Data from Rail

(thousand tonnes)

	RAIL					
	1990 or base year	1991	1992	1993	1994	1995
Austria	707	755	751	775	775	774
Belgium[2]	1 000	1 000	1 000	1 000	1 000	1 000
Canada	6 300	5 800	6 100	6 100	6 300	5 700
Czech Republic	738	601	561	456	441	390
Denmark[1]	483	465	448	452	456	451
Finland	240	--	230	--	290	--
France[1&3]	1 100	1 000	1 000	900	900	800
Germany	3 000	--	--	--	--	--
Hungary	495.3	--	--	313.8	300.8	--
Ireland	114	110	92	92	184	184[1]
Italy[2]	600	--	--	600	600	--
Latvia	881.4	529	456	270	120	107
Lithuania	315	--	--	201	--	--
Netherlands	100	100	100	100	100	100
New Zealand	--	--	--	--	--	--
Norway	--	100	100	100	100[2]	--
Poland	8 960	--	--	--	--	7 327
Russian Federation	18 400	16 700	16 500	16 100	--	--
Slovak Republic	376.9	283.4	233.8	195.6	189.5	--
Slovenia	62	--	--	--	39	41
Sweden	100	--	--	100	--	--
Switzerland	30	30	30	30	30	30
United Kingdom[1]	1 933	1 976	2 049	1 921	1 879	--
United States[2]	31 558.9	28 376.3	30 649.6	28 123.3	29 336.9	29 531.2
European Union[1]	9 096.5	9 440.9	9 074.4	8 928.4	8 446.5	--

Source: Responses to 1996 ECMT questionnaire.   See notes after Table 29.

Table 26. **Annual CO_2 Emissions Data from Shipping**

(thousand tonnes)

	SHIPPING																	
	Internal Navigation						International Marine Bunkers						TOTAL SHIPPING					
	1990	1991	1992	1993	1994	1995	1990	1991	1992	1993	1994	1995	1990	1991	1992	1993	1994	1995
Austria	41	41	41	42	42	43	--	--	--	--	--	--	41	41	41	42	42	43
Belgium[2]													1 000	1 000	1 000	1 000	1 000	1 000
Canada	5 700	6 100	6 100	5 300	5 600	5 300	--	--	--	--	--	--	5 700	6 100	6 100	5 300	5 600	5 300
Czech Republic	54	51	51	48	47	47	--	--	--	--	--	--	54	51	51	48	47	47
Denmark[1]	445	462	456	437	445	443	--	--	--	--	--	--	445	462	456	437	445	443
Finland	--	--	--	--	--	--	1 600	1 600	1 800	2 100	2 300	--	1 600	1 600	1 800	2 100	2 300	--
France[1&4]	--	--	--	--	--	--	8 000	8 300	7 900	7 700	6 900	7 100	8 000	8 300	7 900	7 700	6 900	7 100
Germany	2 000	--	--	--	--	--	8 000[1]	--	--	--	--	--	2 000	--	--	--	--	--
Hungary	26.8	--	--	167.6	252.9	--	--	--	--	--	--	--	26.8	--	--	167.6	252.9	--
Ireland	56	55	49	52	107	104[1]	61	107	54	171	124	363[1]	117	162	103	223	231	467[1]
Italy	1 200	--	--	1 200	1 200	--	8 400	--	--	7 900	7 900	--	9 600	--	--	9 100	9 100	--
Latvia	955	70	57	105	49	43	--	--	--	--	--	--	--	--	--	--	--	--
Lithuania	13	--	--	3	--	--	273	--	--	253	--	--	286	--	--	256	--	--
Netherlands	--	--	--	--	--	--	1 500	1 500	1 500	1 700	1 900	1 800	1 500	1 500	1 500	1 700	1 900	1 800
New Zealand	--	--	--	--	--	--	1 046	918	875	927	1 337	1 135	1 046	918	875	927	1 337	1 135
Norway[3]	--	2 000	2 200	2 300	2 100	--	--	--	--	--	--	--	--	2 000	2 200	2 300	2 100	--
Poland	--	--	--	--	--	--	702[3]	--	--	--	--	--	702[3]	--	--	--	--	588[3]
Russian Federation[1]	--	--	--	--	--	--	--	--	--	--	--	--	18 100	17 100	15 500	13 400	12 100	11 300
Slovak Republic[3]	142.6	122.4	117.8	78.9	95.4	--	--	--	--	--	--	--	142.6	122.4	117.8	78.9	95.4	--
Slovenia	--	--	--	--	--	--	--	--	--	--	--	--	--	--	--	--	--	--
Sweden	--	--	--	--	--	--	2 800	--	--	--	--	--	2 800	--	--	--	--	--
Switzerland	75	75	75	75	75	75	--	--	--	--	--	--	75	75	75	75	75	75
United Kingdom[2]	--	--	--	--	--	--	5 506	6 479	6 373	6 270	5 471	--	5 506	6 479	6 373	6 270	5 471	--
United States[1&2]	41.3[5]	40.4[5]	41.5[5]	40[5]	40.4[5]	40.8[5]	74.3	74.1	78	66.2	65.9	69.2	115.6	114.4	119.5	106.1	106.3	109.9
European Union[1]	21 211	22 365	21 659	21 685	21 918	--	--	--	--	--	--	--	--	--	--	--	--	--

Source: Responses to 1996 ECMT questionnaire. See notes after Table 29.

Table 27. Annual CO$_2$ Emissions Data from Aviation

(thousand tonnes)

	AVIATION																		
	Domestic Aviation						International Aviation Bunkers						TOTAL AVIATION						
	1990	1991	1992	1993	1994	1995	1990	1991	1992	1993	1994	1995	1990	1991	1992	1993	1994	1995
Austria	1 002	1 171	1 527	1 590	1 476	1 396	--	--	--	--	--	--	1 002	1 171	1 527	1 590	1 476	1 396
Belgium[2]													1 000	1 000	1 000	1 000	1 000	1 000
Canada	10 300	9 300	9 400	9 100	9 700	10 500							10 300	9 300	9 400	9 100	9 700	10 500
Czech Republic	294	373	412	444	529	606							294	373	412	444	529	606
Denmark[1]	67	67	67	68	73	75							67	67	67	68	73	75
Finland	--	--	--	220												520	--	--
France[1&4]	--	--	--										12 300	12 100	13 600	13 900	14 500	14 900
Germany	3 000	--	--										3 000					
Hungary	--	--	--										53.2			120.3	110.9	
Ireland	--	--	--			1 147[1]							1 111	1 078	1 078	1 380	1 226	1 147[1]
Italy	1 800	--	--			--							5 800			6 800	7 000	
Latvia	--	--	--			85						85	168	140	113	86	85	85
Lithuania	442	--	--			--							442	71				
Netherlands	--	--	--			--							500	600	600	600	600	600
New Zealand	789	669			456	1 601							2 156	1 974	1 985	2 044	2 303	2 471
Norway[4]	--	1 200			--	--							--	1 200	1 300	1 300	1 500	--
Poland	--	--			--	--							987[3]					566[3]
Russian Federation[2]	--	--			--	--							52 100	54 000	46 400	40 900	31 900	23 200
Slovak Republic[4]													275.9	305.8	396.1	402.3	400	
Slovenia																		
Sweden													1 500					
Switzerland	2 432	2 220	2 582	2 659	2 717	2 921	1 112	1 213	1 038	1 085	1 125	1 130	1 112	1 213	1 038	1 085	1 125	1 130
United Kingdom[3]	--	--	--	--	--	--							2 614	2 537	2 691	2 757	2 876	
United States[1&2]	--	--	--	--	--	--							164.3	168.9	178.3	183	195.2	205.3
European Union[1]	22 155	22 046	22 618	17 754	19 576	--	63 640[2]	61 850[2]	66 777[2]	75 435[2]	77 549[2]		85 794.6	83 895.9	89 394.8	93 188.6	97 124.8	

Source: Responses to 1996 ECMT questionnaire.   See notes after Table 29.

Table 28. Annual CO_2 Emissions Data from Other Categories

(thousand tonnes)

	OTHER					
	1990 or base year	1991	1992	1993	1994	1995
Austria[1]	1 131	1 143	1 167	1 186	1 202	1 215
Finland[1]	2 100	--	--	--	--	--
Hungary[1]	1 653.5	--	--	--	--	--
Italy[3]	2 600	--	--	2 900	2 900	--
Lithuania[1]	(1 293)	--	--	--	--	--
Netherlands[1]	1 800	1 800	1 800	1 800	1 900	1 900
Norway[5]	--	700	700	700	700	--
Slovenia[1]	183	--	--	--	--	86
Switzerland[1]	975	965	960	950	940	940
United States[2&6]	113 970.8	98 636.8	85 161.9	82 646.6	79 819.5	79 100.9
European Union[1&3]	153.7	143.8	145.1	88.9	322.9	--

Source: Responses to 1996 ECMT questionnaire.   See notes after Table 29.

Table 29. Annual CO$_2$ Emissions: Transport Sector and Total

	TRANSPORT SECTOR EMISSIONS (thousand tonnes)						TOTAL EMISSIONS (million tonnes)						PER CENT TRANSPORT EMISSIONS					
	1990*	1991	1992	1993	1994	1995	1990*	1991	1992	1993	1994	1995	1990*	1991	1992	1993	1994	1995
Austria	16 161	17 863	18 235	18 068	18 425	18 712	59.6	61.6	61.6	62	62	62.4	27.1	29.0	29.6	29.1	29.7	30.0
Belgium	22 000	21 500	24 000	25 000	26 000	27 000	110	--	--	--	--	--	20.0	--	--	--	--	--
Canada	140 100	134 500	136 000	139 000	146 700	149 500	451.9	433.9	453.3	463.3	473.2	482.3	31.0	31.0	30.0	30.0	31.0	31.0
Czech Republic	7 926	7 656	8 321	8 977	9 675	9 966	165	163	151.3	147	140	146.5	4.8	4.7	5.5	6.1	6.9	6.8
Denmark	10 236	10 681	10 745	10 708	10 924	10 933	60.2	62.8	59.7	59.5	57.5	57.5	17.0	17.0	18.0	18.0	19.0	19.0
Finland	--	--	--	--	--	--	--	--	--	--	--	--	--	--	--	--	--	--
France[1]	132 800	135 500	139 400	140 400	141 400	143 000	382.7	387.1	391.5	371.4	373.1	376.3	34.7	35.0	35.6	37.8	37.9	38.0
Germany	159 000	162 000	168 000	172 000	169 000	172 000	1 012	975	927	916	897	885	15.7	16.6	18.1	18.8	17.8	19.4
Ireland	6 057	6 252	6 246	7 030	7 136	7 719	31.9	32.9	32	33.5	32.9	35.6	19.0	19.0	19.5	21.0	21.7	21.7
Italy	109 800	--	--	119 100	118 800	--	400.7	--	--	401	396	--	27.4	--	--	29.7	30.0	--
Japan	215 000	224 000	229 000	230 000	242 000	--	1 174.8	1 191.5	1 211.6	1 185.5	1 260.4	--	18.3	18.8	18.9	19.4	19.2	--
Latvia	5 662.5	2 779	2 449	2 172	1 957	2 040	23.6	19.8	16.3	14.5	15	10.7	24.0	14.0	15.0	15.0	13.0	19.0
Lithuania	4 498 (5 791)	--	--	3 044	--	--	36.6	--	--	18.2	--	--	12.3	--	--	16.7	--	--
Netherlands	27 700	28 300	29 200	29 700	30 500	31 700	184.6	188.6	194.6	198	190.6	198.1	15.0	15.0	15.0	15.0	16.0	16.0
New Zealand	11 161	10 972	11 353	11 835	13 056	13 719	27.9	28.1	30	29.6	30.3	29.8	40.0	39.0	38.0	40.0	43.0	46.0
Norway	--	12 100	12 400	13 000	12 900	--	--	33.9	34.4	36	37.6	--	--	35.7	36.0	36.1	34.3	--
Poland	26 065	--	--	--	--	28 498	--	--	--	--	--	--	--	--	--	--	--	--
Russian Federation	234 600	241 600	204 000	187 600	161 500	142 700	2 443.7	2 392	2 241.7	2 287.8	--	--	9.6	10.1	9.1	8.2	--	--
Slovak Republic	5 296.2	4 731.2	4 202.8	4 172.8	4 336.2	--	--	--	--	--	--	--	53.0	--	--	--	--	--
Slovenia	3 192	2 968	3 122	3 699	4 137	4 454	13.6	12.7	12.9	13.2	13.8	14.2	23.4	23.4	24.1	27.9	30.0	31.4
Sweden	20 500	--	--	--	--	--	61.3	--	--	--	--	--	33.4	--	--	--	--	--
Switzerland	14 770	14 855	14 940	15 025	15 215	15 355	44.2	45.3	44.7	43.4	42.7	44.6	33.4	32.8	33.4	34.6	35.6	34.4
United Kingdom	119 744	119 904	121 473	122 481	122 419	--	575	585	565	550	545	--	21.0	21.5	21.5	22.0	22.0	--
United States[1&2]	1 583	1 556	1 580	1 609	1 643	1 682	4 798.8	--	--	--	--	--	32.9	--	--	--	--	--
European Union[1]	755 851	769 586	793 795	807 412	816 622	--	3 220.7	3 264.5	3 188.4	3 134	3 137.5	--	23.5	23.6	24.9	25.8	26.0	--

* Or base year.

Source: Responses to 1996 ECMT questionnaire.

See notes on next page.

Notes to Tables 24, 25, 26, 27, 28 and 29.

Austria
1. Public transport (excluding electric PT), tractors, road works vehicles.

Belgium
1. Includes figures for "Heavy Goods Vehicles and Buses", "Motorcycles and Other" and "Light Commercial Vehicles" categories.
2. Includes Rail, Shipping and Aviation.

Canada
1. Includes vehicles under 600kg and with less than 4 wheels, alternative fuelled vehicles and diesel-powered vehicles used in public administration, industrial and farm sectors.

Czech Republic
1. Includes figures for "Heavy Goods Vehicles and Buses", "Motorcycles and Other" and "Light Commercial Vehicles" categories.

Denmark
1. Base year = 1988.

Finland
1. Non-road mobile machinery.

France
1. Data represent emissions from consumption by the transport sector of fuel sold within metropolitan France (overseas *départements* and *territoires* -- DOM-TOM -- are not included). Data have been corrected for climatic variations.
2. Includes figures for "Heavy Goods Vehicles and Buses" and "Light Commercial Vehicles" categories and also agricultural and military transport emissions.
3. Emissions from electric traction are not included.
4. From French and foreign bunkers in France.

Germany
1. Not included in Totals figure.

Hungary
1. Agricultural vehicles.

Ireland
1. Provisional estimate.

Italy
1. Includes figures for "Private Cars" and "Light Commercial Vehicles" categories.
2. Only includes diesel trains.
3. Includes public administration and utility vehicles.

Lithuania
1. Figures in ( ) represent unspecified data. They are not included in totals.

Netherlands
1. Off road vehicles.

Norway
1. Includes figures for "Private Cars" and "Light Commercial Vehicles" categories.
2. Preliminary data.
3. Not including fishing vessels and mobile oil rigs.
4. Only Norwegian aircraft over Norwegian territory.
5. Motorised tools.

Poland
1. Figures for Motorcycles included in Private Car category.
2. Vehicles up to 3.5 tonnes.
3. Total domestic and international bunkers.

Romania
1. Includes figures for "Heavy Goods Vehicles and Buses" and "Light Commercial Vehicles" categories.
2. Not including military and agricultural transport emissions.

Russian Federation
1. Only includes Russian ships bunkered on Russian territory; Military ships not included.
2. Includes only Russian aircraft bunkered on Russian territory. Military aircraft not included.

Slovak Republic
1. Includes diesel-fuelled agricultural machines, military transport and engine machinery.
2. Includes gasoline-fuelled agricultural and small machinery.
3. From navigation (domestic and international) in the Slovak section of the Danube.
4. From traffic in air corridors over the Slovak Republic.

Slovenia
1. Domestic working machines.

Switzerland
1. Off road transport (agriculture, military, industry).

United Kingdom
1. Emissions based on UNECE source category -- does not include emissions from the generation of electricity to power trains.
2. Includes emissions from fishing, coastal shipping, oil exploration and production, as well as fuel oil use on offshore installations. Marine bunker emissions included only if under 12 miles from shore.
3. Includes only emissions associated with ground movement and take-off and landing cycles up to 1 km from the airport.

United States
1. Preliminary data; emissions figures expressed in million tonnes.
2. Emissions figures have been converted from carbon to CO_2 using conversion factor 44/12.
3. Includes figures for Private Cars and Light Commercial Vehicles categories.
4. Includes freight trucks and buses.
5. Includes recreation boats.
6. Military, lubricant and pipeline NG.

European Union
1. Preliminary data. Source: IEA/OECD, November 1996.
2. International civil aviation.
3. Non-specified (transport).

Table 30. Annual CO$_2$ Emissions Data by Fuel

(thousand tonnes)

	DIESEL											
	Heavy Goods Vehicles						Light Vehicles					
	1990 or base year	1991	1992	1993	1994	1995	1990 or base year	1991	1992	1993	1994	1995
Austria	6 363	--	--	--	--	8 282	--	--	--	--	--	--
Belgium	8 500	8 000	9 000	9 000	10 000	11 000	3 500	4 000	4 000	4 500	5 000	5 000
Canada	17 800	16 900	17 000	18 200	20 500	22 800	3 700	3 800	4 000	4 400	5 000	5 300
Czech Republic	2 395[1]	2 523[1]	2 665[1]	3 313[1]	3 859[1]	3 974[1]						
Denmark	2 089	2 122	2 129	2 042	2 021	2 036	2 227	2 276	2 284	2 309	2 402	2 428
France[28]	39 300[1]	41 400[1]	42 900[1]	43 400[1]	44 200[1]	45 100[1]	14 400[10]	16 200[10]	18 200[10]	20 300[10]	22 900[10]	25 600[10]
Germany	54 400[14]	--	--	--	--	--	[14]					
Italy	402 000	--	--	397 000	395 000	--	19 000[10]	--	--	18 900[10]	17 300[10]	--
Latvia	518	591	479	330	397	353	391	13	10	7	9	8
Lithuania	1 227[1]	1 530[1]	1 033[1]	1 235[1]	1 115[1]	1 042[1]						
Netherlands	7 200[15]	7 400[15]	7 700[15]	7 800[15]	8 000[15]	8 800[15]	6 900	7 000[16]	7 200[16]	7 300[16]	7 500[16]	7 700[16]
New Zealand	412[1]	453[1]	330[1]	323[1]	467[1]	405[1]						
Romania	--	--	--	3 608.8	3 739.3	3 877.4	--					
Russian Federation[5]	56 100[11]	58 100[1]	50 500[1]	42 500[1]	--	--						
Slovak Republic	3 835[3]	3 245[3]	2 487[3]	2 191[3]	2 516[3]	--						
Slovenia	874[1]	798[1]	734[1]	885[1]	1 016[1]	1 101[1]						
Sweden	3 900[1]	--	--	--	--	--						
Switzerland	2 475[12]	2 480[12]	2 485[12]	2 495[12]	2 525[12]	2 545[12]	1 155	1 165	1 185	1 195	1 225	1 245
United Kingdom[5]	33 472	33 605	34 980	37 103	40 580							
United States[18, 19&20]	210.4[21]	204.8[21]	211[21]	222.3[21]	238.6[21]	241.3	13.5	12.5	12.5	12.3	12.2	11.9
European Union[29]	301 738.6 [24]	312 510.1 [24]	323 294.4 [24]	334 622.2 [24]	345 879.9 [24]	--	[24]	[24]	[24]	[24]	[24]	--

164

Table 30 cont.d. Annual CO_2 Emissions Data by Fuel (thousand tonnes)

	PETROL						OTHER (Specify)					
	1990 or base year	1991	1992	1993	1994	1995	1990 or base year	1991	1992	1993	1994	1995
Austria	8 046	--	--	--	--	7 917	214	--	--	--	--	217
Belgium	9 000	8 500	10 000	10 500	10 000	10 000	--	--	--	--	--	--
Canada	80 100	77 400	78 500	80 100	82 400	82 200	16 100[30]	15 200[30]	14 800[30]	15 800[30]	17 100[30]	17 700[30]
Czech Rep.	3 721	3 418	3 654	3 797	3 969	4 188			2[2]	2	4	6
Denmark	4 924	5 290	5 361	5 400	5 528	5 500	--	--	--	--	--	--
Finland	--	--	--	--	--	--						
France[28]	57 800	56 600	55 700	54 100	52 200	49 400	1 100[11]	1 000[11]	1 000[11]	900[11]	900[11]	800[11]
Germany	95 800	--	--	--	--	--	--	--	--	--	--	--
Italy	39 060	--	--	48 500	49 700	--						
Latvia	2 839	1 498	1 436	1 393	726	1 200	1 817	588	476	369	164	145
Lithuania	1 807	1 827	1 212	1 368	1 589	1 728	1 575[7] 352[8] 67[9]	1 515[7] 368[8] 72[9]	175[7] 373[8] 33[9]	104[7] 521[8] 23[9]	112[7] 475[8] 49[9]	106[7] 438[8] 52[9]
Netherlands	10 400	10 600	11 000	11 200	11 500	12 600	3 200[17]	3 300[17]	3 400[17]	3 400[17]	3 500[17]	2 600[17]
New Zealand	5 724	5 729	5 826	5 893	6 082	6 283	1 367[4] 634[4]	1 305[4] 464[4]	1 340[4] 545[4]	1 353[4] 604[4]	1 456[4] 870[4]	1 601[4] 730[4]
Romania	--	--	--	3 223.4	3 791.2	3 997.8						
Russian Federation[5]	89 500	94 300	80 600	72 500	--	--	0.8[6]	1	1	1.1	--	--
Slovak Republic	1 185	1 180	1 320	1 580	1 420	--	276[4]	306	396	402	400	--
Slovenia	2 317	2 169	2 389	2 815	3 122	3 353						
Sweden	12 200											
Switzerland	9 860	9 890	9 920	9 950	10 030	10 100	1 280[13]	1 320[13]	1 350[13]	1 385[13]	1 435[13]	1 465[13]
U. K.[5]	76 219	75 306	75 379	74 429	71 613	--						
U. S.[19 & 20]	939.7[18]	934.3[18]	948.2[18]	978.2[18]	983.8[18]	1 006.9[18]	14.6[22] 1 364[23]	22[22] 1 246.6[23]	29.3[22] 1 151.3[23]	311.6[22] 1 063.3[23]	348.3[22] 1 591.3[23]	388.6 1 448.3
E. U.[29]	357 433[25]	362 142[25]	369 820[25]	368 228[25]	361 632[25]	--	7 228.2[26] 89 566.2[27]	7 033.2[26] 87 991.3[27]	6 672.9[26] 94 088.6[27]	6 894.4[26] 97 711.9[27]	7 242.9[26] 101967.9[27]	--

Notes to Table 30: Annual CO₂ Emissions Data by Fuel

1. Includes HGV and LV.
2. Biogas (Earth gas) for buses in mass urban transport.
3. Includes HGV and LV, Rail and Shipping.
4. Aviation fuel.
5. For road transport only.
6. LPG/CNG.
7. Jet kerosene.
8. Ships oil.
9. LPG.
10. Private diesel cars.
11. Diesel rail.
12. Includes shipping and rail.
13. Kerosene and military fuel use.
14. Figures for LV included in HGV category.
15. Includes figures for internal navigation.
16. Includes off road vehicles.
17. LPG and Kerosene.
18. Emissions figures in million tonnes.
19. Preliminary data; emissions have been converted from carbon to CO_2 using conversion factor 44/12.
20. From Road only.
21. Includes military emissions.
22. CNG.
23. LPG.
24. Light vehicles included in Heavy Goods Vehicles category. All figures include Gas and Diesel oil.
25. Motor petrol.
26. LPG.
27. Includes: Hard Coal, Brown Coal, Other Bituminous Coal and Anthracite, Lignite, Coke Oven Coke, BKB, Natural Gas, Aviation Petrol, Petrol type Jet Fuel, Kerosene type Jet Fuel, Kerosene, Residual Fuel Oil, Combustible Renewables, and Solid Biomass and Animal Products.
28. Data represent emissions from consumption by the transport sector of fuel sold within metropolitan France (overseas *départements* and *territoires* -- DOM-TOM -- are not included). Data have been corrected for climatic variations.
29. Preliminary data. Source: IEA/OECD, November 1996.
30. Includes vehicles under 600kg and with less than 4 wheels, alternative fuelled vehicles and diesel-powered vehicles used in public administration, industrial and farm sectors.

BIBLIOGRAPHY

Belgium (1996), Response to ECMT questionnaire on International Traffic, Ministère des Communications et de l'Infrastructure.

EC (1995), *A Community Strategy to Reduce CO_2 Emissions from Passenger Cars and Improve Fuel Economy.* Communication from the Commission to the Council and the European Parliament. COM (95)689. Commission of the European Communities, Directorate General XI: Environment, Nuclear Safety and Civil Protection, Brussels.

EC- DG VII/DG XIII (1996), *Progress in Telematics Applications for Road Transport in Europe: An Abridged Version of the Report to the High Level Group on Road Transport Telematics.* Brussels: 1996.

EC (Joint Research Centre), June 1996, Final Report Contribution to the STOA project on "The Car of the Future, the Future of the Car", EUR 17277 EN, Institute for Prospective Technological Studies, Seville.

ECMT (1989), *Resolution No. 66 on Transport and the Environment* [CEMT/CM(89)29/Final], European Conference of Ministers of Transport (ECMT), Paris.

ECMT (1991), *Resolution No. 91/5 on the Power and Speed of Vehicles* [CEMT/CM(91)28/Final], European Conference of Ministers of Transport (ECMT), Paris.

ECMT (1993*a*), *Transport Policy and Global Warming,* European Conference of Ministers of Transport (ECMT), Paris.

Specific papers referenced:
Michaelis, Laurie: *Greenhouse Gas Emissions and Road Transport Technology*

ECMT, (1993*b*), *Resolution on Reducing Transport's Contribution to Global Warming* [CEMT/CM (93)21/Final], European Conference of Ministers of Transport (ECMT), Paris.

ECMT (1995), *Dialogue with Vehicle Manufacturers: Declaration on reducing carbon dioxide emissions from passenger vehicles in ECMT countries,* European Conference of Ministers of Transport (ECMT), OECD, Paris.

ECMT/OECD (1995), *Urban Travel and Sustainable Development*, European Conference of Ministers of Transport (ECMT), OECD, Paris.

ECMT (1996), Group on Trends in International Traffic, *Survey of the first draft summaries by country* [CEMT/CS/TTI(96)5], 11 September 1996.

ECMT (1997), Group on Trends in International Traffic and Infrastructure Needs, *Monographs - Spain,* Committee of Deputies, 7 March 1997.

EU Council (1996), *Council Conclusions*: *A Community Strategy to Reduce CO_2 Emissions from Passenger Cars and Improve Fuel Economy* (DG I 8748/96). Brussels: 2 July 1996.

Environment Watch Western Europe, 6 September 1996, *Dutch Announce Array of Measures to Counter Traffic Jams*, p. 5. Cutter Information Corp., Arlington, Ma., United States.

IEA/OECD (1993), *Cars and Climate Change*, International Energy Agency, OECD, Paris.

IEA/OECD (1994), *Climate Change Policy Initiatives, 1994 Update, Volume 1, OECD countries*. International Energy Agency, OECD, Paris.

IEA/OECD (1995), *World Energy Outlook, 1995 Edition*. International Energy Agency, OECD, Paris.

IEA/OECD, (1996*a*), *World Energy Outlook, 1996 Edition*. International Energy Agency, OECD, Paris.

IEA/OECD (1996*b*), *Climate Change Policy Initiatives 1995/96 Update*. Vol. 2 Selected non-IEA Countries. International Energy Agency, OECD, Paris.

IEA, August 1996, Annual data on carbon dioxide emissions from fossil fuel combustion.

Michaelis, L. (1996), Annex 1 Expert Group on the UN FCCC, Policies and Measures for Common Action, Working Paper 1, *Sustainable Transport Policies: CO_2 Emissions from Road Vehicles*, OECD, Paris, July 1996.

Nadis, Steve and MacKenzie, James J. with Ost, Laura, *Car Trouble, A World Resources Institute Guide to the Environment,* Beacon Press, Boston.

NRC (1995) (cited in US DOE 1996). National Research Council, Transportation Research Board. *Expanding Metropolitan Highways: Implications for Air Quality and Energy Use.* Special Report 245. National Academy Press, Washington, D.C.

OECD (1993), *Environmental Performance Reviews: Portugal*, OECD, Paris.

OECD, (1994*a*), *Environmental Performance Reviews: Italy*, OECD, Paris.

OECD (1994*b*), *Environmental Performance Reviews: Japan*, OECD, Paris.

OECD (1995*a*), Environmental Data, Compendium 1995, OECD, Paris.

OECD (1995*b*), *Environmental Performance Reviews: Austria*, OECD, Paris.

OECD (1995*c*), *Environmental Performance Reviews: Netherlands*, OECD, Paris.

OECD (1996*a*), *Main Economic Indicators.* Statistics Directorate, OECD, Paris, November 1996.

OECD (1996*b*), *Short-Term Economic Indicators Transition Economies*, OECD, Paris, February 1996.

OECD (1996*c*), *Environment Performance Reviews: United States*, OECD, Paris.

Solsbery, Lee and Wiederkehr, Peter, *"Voluntary Approaches for Energy-related CO_2 Abatement"*, *The OECD Observer* No. 196, October/November 1995, Paris.

UN-INC (1995), *First Review of Information Communicated by Each Party Included in Annex 1 to the Convention, Compilation and Synthesis of National Communications from Annex 1 Parties, Report by Interim Secretariat,* Intergovernmental Negotiating Committee for a Framework Convention on Climate Change, United Nations, February 1995, Geneva.

UN FCCC (1996), *Ministerial Declaration*, Second Session of the Conference of Parties to the United Nations Framework Convention on Climate Change. Geneva, 18 July 1996.

US DOE (1996), US Department of Energy, Office of Policy and International Affairs, *Policies and Measures for Reducing Energy Related Greenhouse Gas Emissions: Lessons From Recent Literature.* Washington, D.C., July 1996.

US DOT (1993), US Department of Transportation, *Transportation Implications of Telecommuting.* Washington, D.C., April 1993.

Wachs, M. (1994) (cited in US DOE 1996). *"Will Congestion Pricing Ever Be Adopted?"* Access, Spring 1994, no:15-19. University of California Institute for Transportation Studies. Berkeley, California.

Wirth, T. (1996) on behalf of the United States of America: Statement to the Second Conference of the Parties, Framework Convention on Climate Change, Geneva, 17 July 1996.

National Communications to the United Nations Framework Convention on Climate Change

Australia, September 1994: *Australia's National Report under the United Nations Framework Convention on Climate Change.*

Austria, August 1994: Federal Ministry of Environment, *National Climate Report of the Austrian Federal Government*, Vienna.

Belgium, June 1994: *Programme National Belge de Réduction des Emissions de CO_2.*

Bulgaria, 1996: Ministry of Environment, *Bulgaria's First National Communication under the United Nations Framework Convention on Climate Change*, Sofia.

Canada, 1994: *Canada's National Report on Climate Change.*

Confédération suisse, October 1994: *Convention-cadre des Nations Unies sur les changements climatiques, Rapport de la Suisse*, Berne.

Czech Republic, 1994: Ministries of Environment and Industry and Trade, *The Czech Republic's First Communication on the National Process to Comply with the Commitments under the UN Framework Convention on Climate Change.*

Czech Republic, February 1996: Preparatory Committee of the 1996 Regional Conference on Transport and the Environment, *Intentions for Implementation of the Measures to Stabilization and Reduction of the Environmental Burden Caused by Transport in the Czech Republic.*

Denmark, Environmental Protection Agency, Ministry of the Environment: *Climate Protection in Denmark: National Report of the Danish Government in Accordance with Article 12 of the United Nations Framework Convention on Climate Change.*

Denmark, May 1996: Ministry of Transport, *The Danish Government's Action Plan for Reduction of the CO_2 Emissions of the Transport Sector*, Copenhagen.

EC, 1996: *Communication from the Commission under the UN Framework Convention on Climate Change.* Commission of the European Communities, Brussels.

Estonia, 1995. Ministry of Environment, *Estonia's First National Communication under the United Nations Framework Convention on Climate Change*, Tallinn.

Finland, January 1995: *Finland's National Report under the United Nations Framework Convention on Climate Change.*

France, February 1995: *French Republic: National Programme for the Mitigation of Climate Change.*

France, Ministère de l'Environnement and Ademe: *France and the greenhouse effect*: April 1995.

Germany, Federal Environment Ministry, *Environmental Policy,* December 1994.

Greece, February 1995: *Climate Change: The Greek Action Plan for the Abatement of CO_2 and other Greenhouse Gas Emissions.*

Hungarian Commission on Sustainable Development, 1994: *Hungary: Stabilisation of the Greenhouse Gas Emissions: National Communication on the Implementation of Commitments under the United Nations Framework Convention on Climate Change.*

Ireland, October 1994: Department of the Environment, *Ireland: Communication under the UN Framework Convention on Climate Change,* Dublin.

Italy, Ministry of Environment, Ministry for Industry: *National Programme for the Limitation of Carbon Dioxide Emissions to 1990 Levels by the Year 2000.*

Japan, 1994: *Japan's Action Report on Climate Change.*

Latvia, 1995: Ministry of Environmental Protection and Regional Development, *National Communication of the Republic of Latvia under United Nations Framework Convention on Climate Change.* Riga.

Netherlands, August 1994: Ministry of Housing, Spatial Planning and Environment, *Netherlands' National Communication on Climate Change Policies: Prepared of the Conference of the Parties under the Framework Convention on Climate Change.* The Hague.

New Zealand, 1994: *New Zealand's First National Communication under the Framework Convention on Climate Change.*

Norway, 1994: *Greenhouse gas emissions in Norway, Inventories and estimation methods.*

Norway, September 1994: Ministry of Environment, *Norway's National Communication under the Framework Convention on Climate Change.*

Poland, 1994: *National Report to the First Conference of the Parties to the United Nations Framework Convention on Climate Change.* Warsaw.

Portugal, 1994: Ministry of Environment and Natural Resources, *Portuguese Report in accordance with article 12 of the United Nations Framework Convention on Climate Change,* Lisbon.

Romania, January 1995: Ministry of Water, Forests, and Environmental Protection, *First National Communications Concerning the National Process of Applying the Provisions of the Frame Convention of Climatic Changes.*

Slovak Republic, May 1995: *The First National Communication on Climate Change.*

Spain, 1994: Ministerio de Obras Publicas, Transportes y Medio Ambiente, Secretaria de Estado de Medio Ambiente y Vivienda, *Informe de Espana a la Convencion Marco de las Naciones Unidas sobre el Cambio Climatico.*

Sweden, September 1994: Ministry of the Environment and Natural Resources, *Sweden's National Report under the United Nations Framework Convention on Climate Change.*

United Kingdom, January 1994: *Climate Change: The UK Programme, United Kingdom's Report under the Framework Convention on Climate Change*, London.

United Kingdom, December 1995: Department of the Environment, *Climate Change: The UK Programme, Progress Report on Carbon Dioxide Emissions*, Central Office of Information. Great Britain.

United States of America, October 1993: *The Climate Change Action Plan*.

MONITORING OF FUEL CONSUMPTION AND CO_2 EMISSIONS OF NEW CARS

TABLE OF CONTENTS

A. BACKGROUND 177

B. TRENDS 179

    Industry policy 179
    Current and upcoming developments 179
    Historical trends analysis 180

C. MONITORING METHODOLOGY 187

    The concept of a monitoring system 187
    Official data 187
    Commercial and industry association data 189
    The influence of the Directive 93/116/EEC 191
    Additional issues 193

D. DATA 195

NOTES 202

A. BACKGROUND

In their 1995 joint Declaration on reducing carbon dioxide (CO_2) emissions from passenger vehicles in ECMT countries, Ministers of Transport agreed with the vehicle manufacturing industry, represented by the presidents of OICA and ACEA, on a number of joint actions. The main objectives of the Declaration are:
- to substantially and continuously reduce the fuel consumption of new cars sold in ECMT countries;
- to manage vehicle use so as to achieve tangible and steady reductions in their total CO_2 emissions.

Governments and Industry agreed to establish an appropriate system to monitor progress toward these goals.

It was agreed that the monitoring system should analyse trends in the projected specific fuel consumption and/or CO_2 emissions of new cars sold in ECMT Member countries, with data provided annually for each ECMT Member country in a standardised form from 1996. This data would cover (i) the number of new car registrations, making appropriate distinctions among vehicle characteristics; and (ii) the specific fuel consumption (in litres/100 kilometres) and/or CO_2 emissions (in grammes per kilometre) of these cars.

A pragmatic, cost-effective approach to monitoring was adopted. Following an examination of possible existing sources of data, in terms of scope and quality, ACEA and OICA undertook to provide data drawn from an existing high quality industry database. This first report under the Declaration presents monitoring data and analysis for the period 1980 to 1995.

In a related development, in 1996, the European Commission entered into discussions with EU Member States regarding the provision of official data for detailed CO_2 monitoring arrangements. It is envisaged that these arrangements will require Member States to make firm commitments to put in place systems that produce accurate and comprehensive data at some point in the future.

The joint Declaration requires ECMT to provide Ministers with a further progress report in 1999. If possible, that report will utilise the new EU monitoring data. However, if development of the EU system should fall substantially behind schedule, the present monitoring exercise will be updated. This will ensure that there is no duplication of efforts but that monitoring can continue even in the event of the EU system suffering delays.

B. TRENDS

Industry policy

Bringing to market fuel-efficient cars has always been a major priority for Europe's automotive manufacturers. Indeed, with high fuel taxes and prices across Europe, viable developments in this field can give products a much valued competitive-edge in the highly competitive European marketplace. Of course, the industry has also to produce cars that are safe, reliable and eco-friendly, that meet the consumers' requirements and transportation needs, and all at a price car-buyers can afford.

The industry's keen interest in, and responsible attitude towards, fuel efficiency is further evidenced by the Joint Declaration. In support of this Declaration a number of joint industry-government initiatives are now being actively and co-operatively pursued, including this monitoring exercise.

Other examples of the automotive industry's responsible attitude are evident by commitments made at national and EU-Level:

- the German car industry has committed to cut the average fuel consumption of the cars sold in Germany by 25% over the 1990 to 2005 period;
- French manufacturers have committed themselves to reduce the average CO_2 emissions of their cars sold in France to 150 g/km by 2005.
- Volvo has announced that it will reduce the average fuel consumption of its cars sold in the EU by 25% from 1990 to 2005;

Moreover, since September 1996, ACEA (the European Automobile Manufacturers Association) has been in deep, constructive and co-operative discussions with the European Commission to lay the basis for an agreement on a reduction of CO_2 emissions by passenger cars.

Current and upcoming developments

Reductions in the average fuel consumption of new vehicles purchased can arise in two ways: through technical improvements in the new models marketed by manufacturers; or through consumers choosing lower consumption vehicles from among existing models (e.g. through down-sizing). Both phenomena have contributed to improved fuel efficiency in the last fifteen years.

To improve fuel economy and reduce CO_2 emissions, since 1996 car manufacturers have been developing, evaluating and testing an array of technical improvements that could potentially and progressively become available and affordable in the 2000 to 2010 period. This is the time-frame in which major new technical developments can realistically be expected to be incorporated into new car/engine programmes. Improvements relate to both powertrain and vehicles, with approaches relating to weight reduction, greater core engine efficiencies, reduced frictional losses, reduced rolling resistance, improved aerodynamics and more efficient transmission systems. Specific examples include:

- continued development of partial lean-burn, spark ignition technologies;
- small high-rated pressure charged and intercooled engines;
- multivalve and variable valve timing technologies;
- variable geometry turbo technology;
- two-phase high temperature cooling;
- reduced tyre and brake losses;

- low friction lube oils/ reduced engine friction losses;
- reduced ancillary power losses such as through electric power steering;
- increased use of lightweight materials such as light alloys and composites for powertrain and body parts;
- optimised Continuously Variable Transmissions;
- further development of direct injection and multivalve engines;
- manual 6 speed gearboxes, robotised mechanical gearboxes and 5 or 6 speed automatics (additional forward gears permit improved gear ratios providing a potential for improved fuel efficiency).

The efforts of car manufacturers to reduce CO_2 emissions are focused on the development of technical improvements that are both available and affordable. At the same time they are continuing to pursue R&D programmes to achieve technological breakthroughs. Examples of possible breakthroughs include:

- development of de-NO_x catalyst technologies to enable direct injection 'full lean burn' petrol engines (dependent on improved fuel quality – low sulphur fuel);
- modified Miller cycle engine technology;
- infinitely variable transmissions;
- energy recovery systems/regenerative braking;
- advanced battery technology;
- fuel cells;
- hybrid cars.

If a specific technological breakthrough were to be achieved and the resultant technical development became affordable, as well as available, then the potential for CO_2 emissions reductions would be further improved. Adoption into the marketplace would, of course, only be gradual. The manufacturers consider that the pursuit of technological breakthrough possibilities would be enhanced by properly focused national and European automotive R&D programmes backed by appropriate levels of public funding (the EU 'Car of Tomorrow' initiative could provide an appropriate delivery mechanism, if appropriately funded).

As in the past (see Historical Analysis below), the fuel economy improvement effects arising from technical improvements brought to market by car manufacturers will be reduced by the adverse impact of the up-coming auto-regulatory agenda and by other essential product improvements introduced to address societal needs, such as vision enhancement (larger glazed surfaces) or ergonomic features to meet the needs of an ageing population, a taller population and so on. Consumer demand for accessories such as air conditioning also has an impact. Although air conditioning is turned off in the fuel consumption test cycle the weight of the unit has an effect.

Car manufacturers consider a wide-range of factors can contribute to an effective car CO_2 reduction strategy – not just technical improvements to the vehicle itself. For example they are highly supportive of initiatives relating to:

- telematics and infrastructure improvements;
- labelling and providing adequate consumer information;
- scrappage schemes that remove older inefficient cars from the park;
- the establishment of an alternative fuels infrastructure;
- training schemes to encourage more fuel-efficient driving behaviour;
- improved maintenance and inspection of vehicles;
- the establishment of fuel standards that enable cars to achieve maximum combustion efficiency.

The industry also believes that the establishment of other vehicle-related environmental, safety and recycling objectives/regulations should take account of fuel efficiency impacts.

Historical trends analysis

Monitoring under the Declaration is based on 7 national markets (figure 1). This covers 71.4% of the new cars registered in all ECMT countries (data from all 15 countries covered by CCFA data could increase the share up to 87%, see table D1). Data for 1995 is included despite the inaccuracies resulting from the introduction of the new test cycle, which for the year 1995 are estimated to be small.

◆ Figure 1. **First registrations**

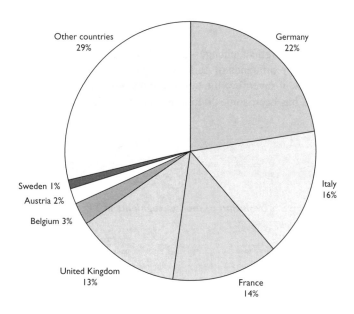

Source: CCFA/AAA, 1996.

◆ Figure 2. **Weighted average fuel consumption, all new cars**
(litres per 100 km)

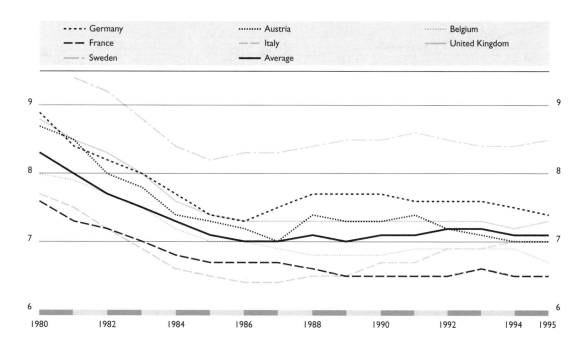

Source: ACEA/OICA, 1997.

The average fuel consumption of the new car fleet in Europe has decreased by almost 15% over the 1980 to 1995 period (figure 2). As can be seen in table 1, average consumption was 8.3 litres per 100 km in 1980 and in 1995 stood at 7.1 litres per 100 km. In more detail, specific fuel consumption fell 15% in the 5 years from 1980 to 1985, since when there has been no further fall and in most markets a small rise in consumption. Part of these trends is explained by changes in the real price of fuel (see figure 5), allowing for a lagged response, and the cost of motoring relative to incomes.

Average fuel consumption trends reflect changes in market segmentation and the penetration of diesel models in the market, together with the influence of parameters such as weight, engine displacement and power. The following figures describe trends in several of the key variables. As figure 3 shows, the small car segment of the market has expanded and there has been some contraction in the medium and larger car segments (see also table D3).

◆ Figure 3. **Evolution of market segments in 7 countries**

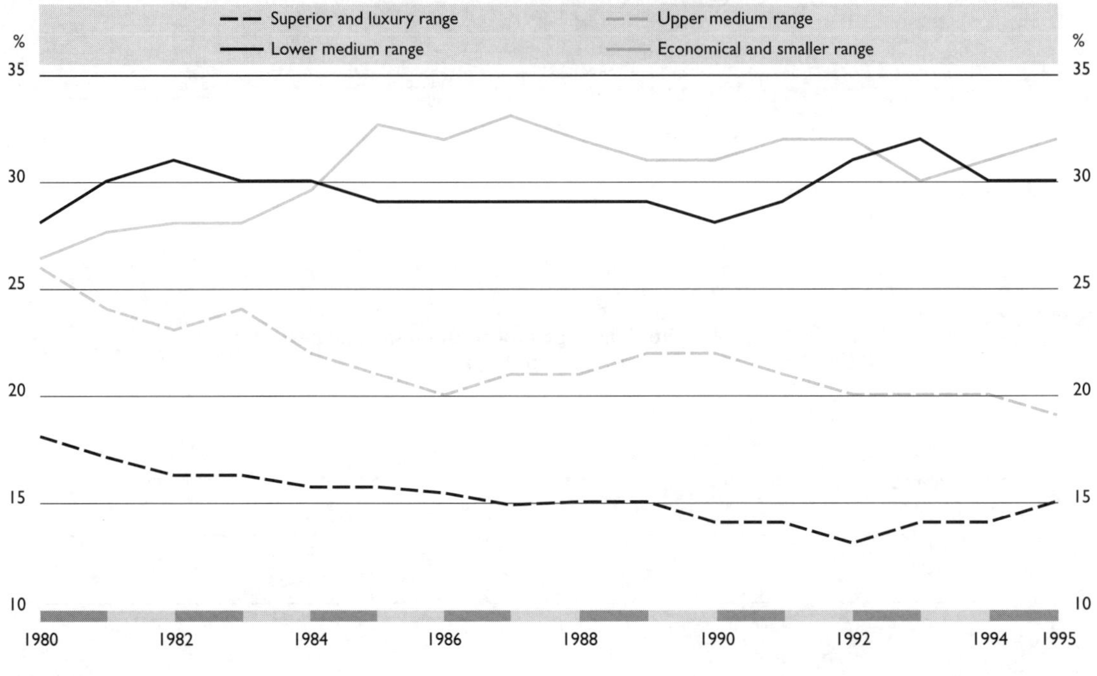

Source: ACEA/OICA, 1997.

With benefits to both fuel economy and CO_2 emissions, diesel penetration has increased from 7% in 1980 to 22% in 1995 (see figure 4 and table D2)[1]. Together with improvements in the quality of diesel engines, this trend has been influenced to a large degree by differences between the rates of excise duty imposed on diesel and petrol (see figure 5). Technically, the fuel efficiency of petrol and diesel engined cars have improved by similar amounts since 1980, specific diesel consumption has been reduced 15% on average and petrol consumption 13%. Combining technical and market factors, it is estimated that diesel cars have contributed between 2 and 3 percentage points to the 15% improvement in overall fuel economy since 1980, around 1.5% accounted for by market penetration and around 10% from technical improvements.

Table 1. **Diesel and Petrol Engined Vehicle Estimated Specific Fuel Consumption**

Vehicle type	1980		1995	
	Estimated average fuel consumption	Market share (7 countries)	Estimated average fuel consumption	Market share (7 countries)
Petrol	8.4 l/100 km	93%	7.3 l/100 km	78%
Diesel	7.4 l/100 km	7%	6.3 l/100 km	22%
Total	8.3 l/100 km		7.1 l/100 km	

Source: ACEA/OICA, 1997.

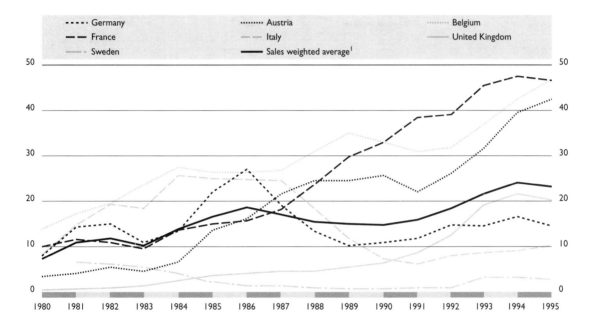

◆ Figure 4. ***Penetration of diesel cars***
(% of new sales)

1. All Europe for which data recorded (8 countries in 1980, 17 countries in 1995).
Source: ACEA/OICA, 1997.

Figure 8 gives more details in two markets, Austria (fuel consumption of petrol and diesel cars, and diesel penetration) and Switzerland (fuel consumption and weight). All these parameters exist in the CCFA database and it is (theoretically) possible to produce all the necessary figures, i.e. the development of the fuel consumption in the different segments as well as by market shares.

The data on specific fuel consumption do not fully reflect the real improvements which have occurred in new car efficiency. These improvements are hidden by the fact that there have been major changes in the actual new cars sold in Europe through the 1980s and into the 1990s. These changes have arisen for a variety of reasons – most notably tougher automotive regulations, demographic and social changes and customer demands. Although a few of these changes have had a positive effect on fuel economy, the majority have had significant negative impacts.

Automotive regulations relating to emissions, safety, noise and so on have all been tightened, with consequent adverse implications for new car fuel economy. Hypothetically, adding a three way catalytic converter to a vehicle equipped with technologies such as an engine management system, electronic fuel injection and multi-valve engine, results in a fuel economy penalty of 5% on average for new cars, according to industry estimates.

◆ Figure 5. **Petrol and diesel pump prices – including excise taxes and VAT**
(1990 US$/l)

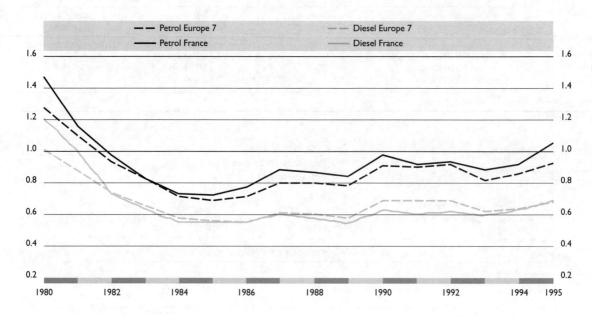

Source: International Energy Agency database, Paris, 1997.

◆ Figure 6. **Sales weighted average power and engine capacity, 7 countries**

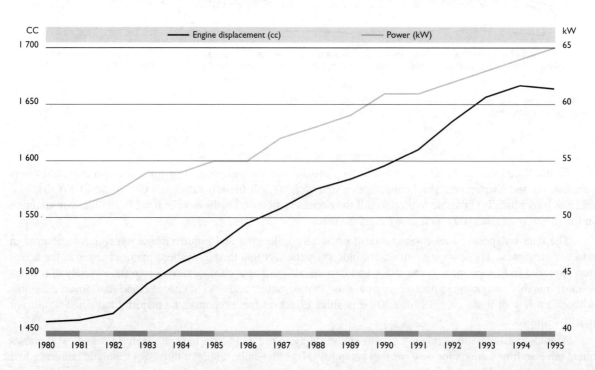

Source: ACEA/OICA, 1997.

◆ Figure 7. *Sales weighted average power*

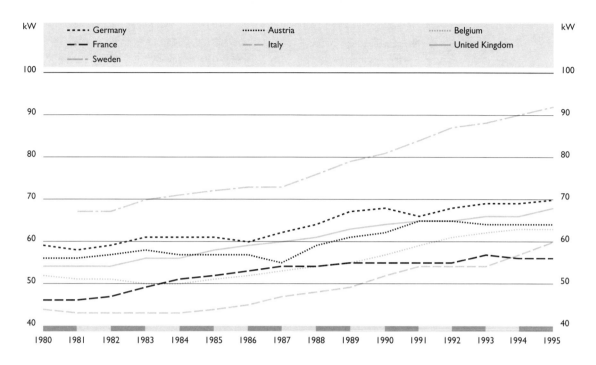

Source: ACEA/OICA, 1997.

Social demands, for example to cope with safety and security concerns or with demographic changes, have given rise to essential product developments or have imposed design constraints that have also been detrimental to fuel economy. The increased demand for power steering is an example.

Driven by customer needs and regulatory pressures the trend has been towards heavier vehicles with higher cubic capacity engines and increased engine power (see figures 6 and 7 and tables D4 and D5). Car buyers have also sought additional product features to enhance driveability, safety, comfort and quality.

All product additions, whether to meet regulatory requirements or market demands add to the weight of vehicles and have other knock-on effects on vehicle structures, that adversely impact fuel economy. In fact over the 1980 to 1993 period the curb weight of the average new car in Europe has risen by 134 kg – from 944 kg to 1 078 kg (an increase of over 14%). In general new models are heavier than the product they replace; this is well illustrated by looking at the curb weight of three generations of the base-model VW Golf:

– Golf I (1974): 750 kg
– Golf II (1983): 845 kg
– Golf III (1991): 960 kg

As an illustration of the main detrimental effects of regulations, essential product improvements and consumer demand, the European auto industry has undertaken a comparison of two similar products produced by one of its member companies. One model was produced in 1984, the other in 1996. The analysis suggests that emissions, safety and noise regulations, together with product improvements to meet social demands, led to an increase in fuel consumption of over 15%. This increase can be attributed as follows:

4.9% to emissions regulations;
7.3% to product improvements associated with safety and noise;
2.8% to social demands such as the inclusion of power steering.

Technical improvements introduced by the manufacturer more than compensated. From the higher fuel consumption level, these technical improvements lowered consumption by some 23%, with a net reduction in specific fuel consumption of 8% between the two models.

◆ Figure 8. **Fuel consumption and diesel penetration in Austria and Switzerland**

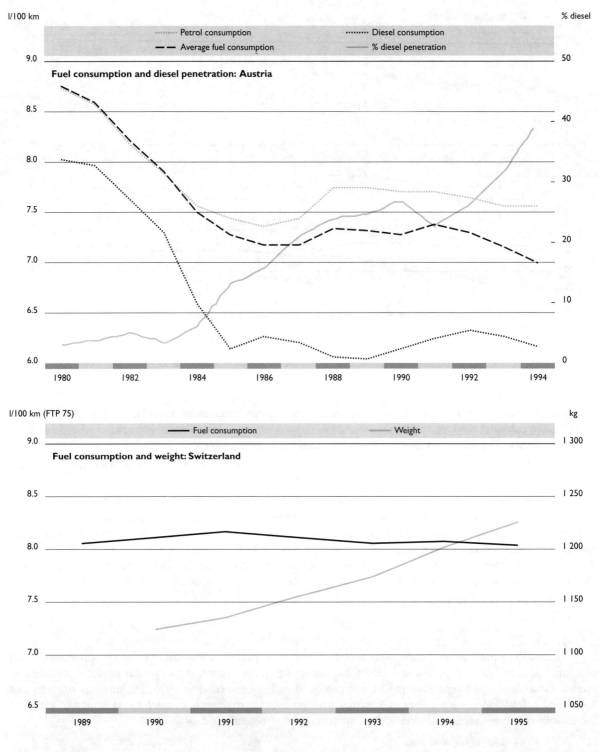

Source: INFRAS, 1996.

C. MONITORING METHODOLOGY

The concept of a monitoring system

Conceptually, the design of a monitoring system is straightforward. Any monitoring system requires basically two pieces of information:
- fuel consumption for the different models of vehicle (measured in a standardised way);
- and the number of cars (by model) sold and registered in a particular time unit in a particular area (country).

The first item is generally part of the type approval data set registered for each new model of car, the second corresponds to annual vehicle registrations.

The main problem in implementing a monitoring system at a European or ECMT level is that the procedures for type approval and registration follow national rules, and the procedures followed in prescribing and collecting data differ from country to country. In addition, even at the national level, creating a relationship between the two data sets (type approval data and registration data) in such a way that the weighting procedure yields a reliable figure of the overall average energy consumption of all cars presents major difficulties.

Due to these difficulties, no official monitoring systems so far exist, even at national level. However, several private organisations have undertaken to produce average figures for fuel consumption, in general at national level. For example:

- The VDA (Verband Deutscher Automobilindustrie) has regularly published sales-weighted figures of the fuel consumption of the German fleet (of German car industry products) since 1978.
- The VDIK (Verband Deutscher Importeure von Kraftfahrzeugen) has produced corresponding figures for imported cars in Germany since 1978.
- Similarly, VSAI/AISA (the Swiss Association of Importers) publishes data on the Swiss market.
- On an ad hoc-basis different institutions (e.g. research institutes) derive figures for specific applications, e.g. the Technical University of Graz for Austria.

The most comprehensive data, however, is contained in the BDSA (Base de Données Statistiques Automobiles) data base maintained by the AAA (Association Auxiliaire Automobile) under the umbrella of the CCFA (Comité des Constructeurs Français d'Automobiles). This database allows production of sales-weighted figures for the fuel consumption of cars in 15 countries.

Official data

Official data relevant for monitoring is contained in three kinds of document. The European Union issues type approval certificates for all vehicles markets in the Union, this includes data on specific fuel consumption. National governments issue national type approval documents which in most cases are based in part on EU type approval certification. Finally, local or national registration agencies maintain data on the number of new car registrations. Only in a few cases are these the same agency that issues national type approval certification. The data recorded on the registration certificate and associated records varies from country to country and may or may not include specific fuel consumption. Figure 9 describes the procedures by which the necessary official data are (or could be) collected and compiled.

◆ Figure 9. *Sources for official data on fuel consumption and car registrations*

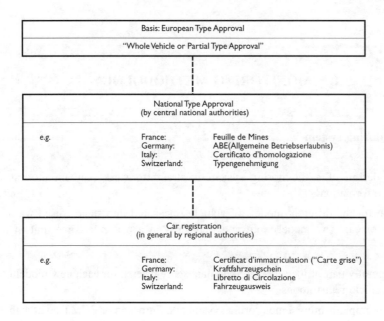

Manufacturers can issue "certificates of conformity" (COC) with EU type approval certification for the individual vehicles that they sell. However they do not issue such certificates on a routine basis, and only do so where national authorities require a copy. As part of the process of market liberalisation within the European Union, amendments to the EU Directives under which COCs are issued (70/156/EEC and 92/53/EEC) are development by the Commission. Proposals include making compulsory the issue of COCs to car buyers and adding a record of specific fuel consumption to the certificate. This would greatly facilitate monitoring but at present few national homologation or registration authorities make use of the COC system.

The degree to which source data can be qualified as "official" is an issue in some quarters in so far as data is seen to be objective. Sufficient transparency in collection and compilation procedures should, however, be able to render any data source acceptable.

Fuel consumption data

Specific fuel consumption is recorded as part of the (EU) type approval certificate. This means that, in principle, the necessary data on fuel consumption is available.

Changes in methodology, however, have resulted in a break in time series data. In the past, fuel consumption in the EU was measured according to Directive 80/1268. Since 1991, a new test cycle for measuring emissions has been introduced (the new European test cycle) and in order to reduce discrepancies between emissions and fuel consumption, the Directive on fuel consumption was revised and published as Directive 93/116/EEC. This Directive has been in force since 1 January 1996 for all new type approvals, and from 1 January 1997 for all newly registered cars.

Under the previous Directive it was left to the Member States whether fuel consumption was recorded as part of the national type approval or not. Under the new Directive all Member states will be obliged to adopt fuel consumption as part of their national type approval schemes (whether or not this information is included in vehicle registration documents is optional).

A major problem arises due to the fact that national type approval procedures are very heterogeneous. Most countries define some sort of "type homologation" (in France e.g. *Feuille de Mines*, in Germany ABE, in Italy *Certificato d'homologazione*, etc.), usually based on the EU type approval certification. According to national rules, different versions and variants of a model can be pulled together in one "type". The level of differentiation varies from country to country. While Germany, for example, uses a very differentiated segmentation, most other countries apply some sort of aggregation. This means that several versions with different fuel consumption might be lumped together under one "type approval number".

An example in Switzerland[2] shows that, theoretically, 15 versions (of the same type and motor) could receive the same type approval number. The fuel consumption of these 15 versions, however, varies by some 10 % (in the example chosen from 8.2 up to 9.3 l/100 km). Thus attribution of the correct fuel consumption to a particular "type" is a major problem.

Registration data

While the type approval data varied significantly in the past, it is likely that in the future the system will be harmonised. The systems of registration though are much more heterogeneous between the different countries. While some countries do not record the number of newly registered cars with any details at all (e.g. Portugal, France) other countries collect and compile this information. The crucial question, however, is whether the differentiation of the recorded registration data is detailed enough to allow an accurate reference to the type approval data to be made.

Example 1: In Switzerland all cars are registered according to type approval reference number. However, as shown above, this information is not detailed enough to make a precise reference to the relevant fuel consumption data since several versions of the same car can carry the same type approval number. It would therefore require additional information (or assumptions) to identify the correct version. Furthermore, according to Swiss officials, the tendency is rather to *decrease* than to *increase* the number of type approval categories, which means that the information collected is tending to become more and more aggregate in nature.

Example 2: In Germany no link can easily be made from registration data to fuel consumption data. All technical data of the different car models are registered in a special reference data set (*Datenblattdatei*) whose primary key is defined by the three dimensions producer / type / version[3]. For the registration process this information is available to all registration authorities. Therefore, it is theoretically possible to set up a link between registration and technical data. However, fuel consumption is not contained in the reference data set (Datenblattdatei). According to German officials, it would be a costly task to complement the technical data set with fuel consumption.

In most other countries similar difficulties will be encountered. Therefore, if a monitoring system at a very disaggregate level should be installed, a considerable effort would be necessary to harmonise the national approval and particularly the registration procedures.

Commercial and industry association data

General remarks

Several private organisations / associations (like VDA, VDIK, VSAI etc.) publish figures which correspond to the desired result of a monitoring system, i.e. average sales-weighted fuel consumption. In general, these figures are based on data recorded by manufacturers or importers and then compiled by the corresponding associations. This means that each manufacturer or importer has to go through the same exercise in linking sales figures by model/type and fuel consumption. Since each manufacturer or importer just has to deal with its own products (models/types/versions), it can be assumed that this procedure can be performed with a high level of competence and reliability.

For a monitoring system, the requirement remains that this process be made transparent. This might imply that certain details are made accessible to third parties. This, on the other hand, might touch upon commercial interests. Another problem arises from the fact that not all producers or importers are members of the corresponding associations. Therefore, the associations do not necessarily cover 100% of the market.

The CCFA database

The most comprehensive database encountered is the BDSA (Base de Données Statistiques Automobiles) database maintained by the AAA / CCFA (*Comité des Constructeurs Français d'Automobiles*). This database allows production of sales- weighted figures of specific fuel consumption for passenger cars in 15 countries. These 15 countries cover about 87 % [1994] of all first car registrations in the ECMT countries (see table 2)[4] and might therefore be considered to be representative for the ECMT countries.

None of the 15 countries covered are in eastern Europe. As car sales grow in this region it will become important to cover the larger markets, and the CCFA database is expected to evolve in this direction.

Table 2. **Passenger cars in use and first registrations of the ECMT countries**

Country	passenger cars in use in [1000]; 1.1.94	first registrations in [1000]; 1990	first registrations in [1000]; 1994	Source	covered by CCFA
Austria	3 367.6	288.6	273.7	1)	x
Belgium	4 210	473.5	404	1)	x
Bosnia-Herzeg.					
Bulgaria	1 443	33.4	76.5	1)	
Croatia	646	66.2	58.5	1)	
Czech Republic	2 694	107.6	256.3	1)	
Denmark	1 675	80.8	139.3	1)	x
Estonia	317		51.2	1)	
Finland	1 873	139.1	67.2	1)	–
France	24 385	2 309.1	1 972.9	1)	x
Germany	39 202	3 041	3 209	1)	x
Greece	1 959	115.4	118.4	1)	–
Hungary	2 092	83.9	199.3	1)	
Ireland	891	81.2	80.4	2)	x
Italy	29 600	2 096	2 319	1)	x
Latvia	391	10.6	5.9	3)	
Lithuania	598				
Luxembourg	209	34.6	29.9	1)	x
Moldova	166	13.1	1	3)	
Netherlands	5 820	502.7	433.9	1)	x
Norway	1 653.7	63.7	91.3	1)	x
Poland	6 771	358.1	486.6	1)	
Portugal	2 210	211.1	243.3	1)	x
Romania	1 793	100	227		
Slovak Republic	995	49.5	30	4)	
Slovenia	632.6	75.5	47.2	1)	
Spain	13 441	982.1	939	1)	x
Sweden	3 566	229.9	156.4	5)	x
Switzerland	3 138	323	265.9	1)	x
Turkey	3 218	267.8	251.3	1)	
United Kingdom	23 832	2 008.9	1 910.9	1)	x
Total	**182 788.9**	**14 146.4**	**14 345.3**		
% covered by CCFA	90.0 %	86.9 %			

Sources:
1. International Road Federation, Geneva
2. Society of the Irish Motor Industry (SIMI), Dublin
3. International Road Federation, Geneva (only imports)
4. Automotive Industry Association, Prague (1990: including Slovakia)
5. Association of the Swedish Motor Industry (BIL), Stockholm

CCFA obtains *registration data* from different sources (industry associations, national statistics offices etc., see table 3). The *fuel consumption* data are derived in general from type approval data complemented by ad hoc inquiries at manufacturers and other sources. According to CCFA, the relationship between registrations and fuel consumption is based on type approval certification and the Vehicle Identification Number (VIN)[5]. The data can thus be broken down by all relevant parameters, country, manufacturer, model/type, homologation type, fuel type, weight, engine size, etc. It is therefore in principle possible to derive the fuel consumption as well as market shares for all different levels of aggregation.

Table 3. **Sources of registration data of the CCFA Database**

Country	Source
EU	
Austria	OSZ
Belgium	FEBIAC
Denmark	AIS
Finland	AUTOTUOJAT
France	AAA
Germany	KBA/VDA
Greece	ASSOCIATION OF MOTOR VEHICLES IMPORTERS
Ireland	SIMI
Italy	ANFIA/UNRAE
Luxembourg	SOCIETE NATIONALE DE CONTROLE TECHNIQUE
Netherlands	RAI
Portugal	ACAP
Spain	ANFAC
Sweden	BRANSCHDATA/BIL
United Kingdom	SMMT
OUTSIDE EU	
Norway	OFV
Switzerland	OFFICE FEDERAL DE LA STATISTIQUE

A number of assumptions and approximations have to be made by CCFA since several versions of a car carrying the same identification number, with different fuel consumption, might exist (see above). For CCFA data to be recognised "officially" it would be indispensable for CCFA to publish the detailed assumptions and rules which are applied for each country. For instance in the Swiss case, according to Swiss officials, CCFA does *not* get the VIN code together with the registration data. So CCFA most likely has to consult additional sources and apply specific rules in order to define a relationship to fuel consumption. Nevertheless, CCFA data has demonstrated an adequate level of accuracy. ACEA members were given the opportunity by CCFA to examine sample extracts from their database and only a few minor discrepancies with companies' own data were reported.

In order to check the accuracy and the compatibility of different data sources, figure 10 shows a comparison of the fuel consumption of the new cars in Germany and Austria. The data were derived on the one hand from a table presented by ACEA and drawn from the CCFA database[6] and on the other hand from German sources (VDA and VDIK)[7] and the TU Graz (for Austria).

The differences in the German data are almost negligible. This good correspondence might partially be due to the fact that similar data sources were used. Therefore, the comparison of the Austrian data is of greater relevance. Also in that case, the correspondence in general can be judged to be satisfactory.

The influence of the Directive 93/116/EEC

Directive 93/116/EEC has been in force since 1/1/1996 for all new type approvals and from 1/1/1997 for all newly registered cars. This implies that the fuel consumption data recorded for an individual car will change significantly. Due to this incompatibility of the measuring systems it is difficult to create a precise time series including past as well as future developments. In particular, CCFA has not yet integrated fuel consumption figures according to the new test cycle into its database. However, a number of organisations have made test measurements that enable a comparison of results from the two test cycles. The German car manufacturers, for instance, agreed upon publishing the values according to the old *and* the new system for the years 1996, 1997 and 1998. This should allow, at least in the German market, for continuous time series data. At the same time, a qualitative indication of developments during the transition period will be provided.

There is no broad empirical basis available yet which would allow derivation of a precise correlation between the fuel consumption based on the old and new cycles. It can be assumed that each car reacts in different ways to the change in test cycle. In general, one has to expect that the new cycle, which includes for the first time

◆ Figure 10. **Comparison of fuel consumption data derived from different sources**

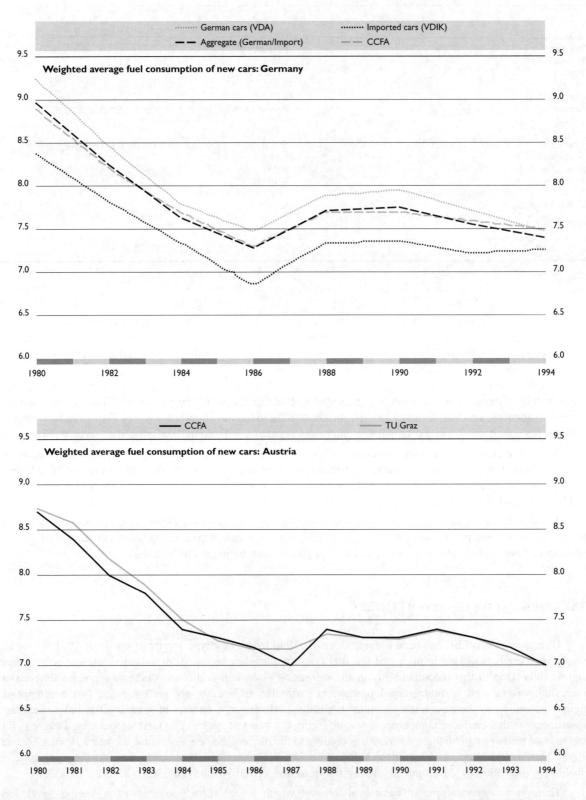

Source: INFRAS, 1996.

◆ Figure 11. **The new test cycle according to Directive 93/116/EEC tends to increase the values for fuel consumption compared to the test procedure according to Directive 80/1268**

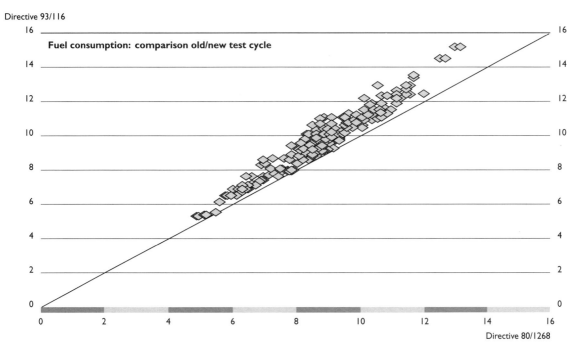

Source: INFRAS, 1996.

a cold start period, tends to produce higher values for fuel consumption (of the order or magnitude of 10 %), and that the higher the fuel consumption of a car the bigger the difference (in absolute terms). With the new test cycle the figures for fuel consumption will move closer to the actual fuel consumption of vehicles in use. Figure 11 shows the correlation between the old and the new values for a sample of cars of three manufacturers.

Additional issues

In addition to the aspects discussed, there remain other issues to be addressed when implementing a monitoring system for specific fuel consumption. The assumptions made with respect to these items must be made transparent for detailed monitoring of specific fuel consumption.
- Definition of passenger cars *versus* light duty trucks:
  The borderline between the definition of passenger cars and light duty trucks may vary between countries. In some countries, for instance, for fiscal reasons some models tend to be registered as light trucks rather than as cars although their function is clearly passenger transport. CCFA generally takes over the definitions of the source of the particular country.
- Off road vehicles:
  Some publications on development efficiency explicitly exclude off-road vehicles although their usage increasingly approaches that of conventional passenger cars.
- Additional features (e.g. air conditioning):
  Air conditioning equipment tends to increase fuel consumption by 0.5 to 1 l / 100 km when running. However, the fuel consumption registered in the official type approval certifications do not take into account this excess consumption since according to the regulation the fuel measurements are performed without these devices. The use of such accessories is a source of significant difference between test cycle and on road emissions, and may warrant quantitative, or at least qualitative examination, in the monitoring system.

D. DATA

Most of the following data tables were provided from the CCFA database by ACEA and OICA in January 1997. With presentation of this report to ECMT Ministers they enter the public domain.

Table D1. **Weighted average fuel consumption, all new cars** (litres per 100 km)

	1980	1981	1982	1983	1984	1985	1986	1987	1988	1989	1990	1991	1992	1993	1994	1995
Germany	8.9	8.4	8.2	8	7.7	7.4	7.3	7.5	7.7	7.7	7.7	7.6	7.6	7.6	7.5	7.4
Austria	8.7	8.5	8	7.8	7.4	7.3	7.2	7	7.4	7.3	7.3	7.4	7.2	7.1	7	7
Belgium	8	7.9	7.7	7.5	7.2	7	7	6.9	6.7	6.8	6.8	6.9	6.9	6.9	6.9	6.7
France	7.6	7.3	7.2	7	6.8	6.7	6.7	6.7	6.6	6.5	6.5	6.5	6.5	6.6	6.6	6.5
Italy	7.7	7.5	7.2	6.9	6.6	6.5	6.4	6.4	6.5	6.5	6.7	6.7	6.9	6.9	7	7
UK	8.8	8.5	8.3	8	7.6	7.4	7.3	7.3	7.3	7.3	7.3	7.3	7.3	7.3	7.2	7.3
Sweden		9.4	9.2	8.8	8.4	8.2	8.3	8.3	8.4	8.5	8.5	8.6	8.5	8.4	8.4	8.5
Average		8	7.7	7.5	7.3	7.1	7	7	7.1	7	7.1	7.1	7.2	7.2	7.1	7.1

Source: ACEA/OICA, 1997.

Table D2. **Penetration of diesel cars** (% of new sales)

	1980	1981	1982	1983	1984	1985	1986	1987	1988	1989	1990	1991	1992	1993	1994	1995
Germany	8	14.3	15	11	13.3	22.1	27.1	19.2	13.4	10.2	10.8	11.8	14.8	14.6	16.6	14.6
Austria	3.3	4.2	5.5	4.6	6.7	13.7	16.2	21.7	24.6	24.6	25.7	22.1	26.2	31.6	39.6	42.6
Belgium	13.8	17.3	19.5	23.7	27.5	26.4	26.3	26.9	31	35.1	32.9	30.8	31.8	36.9	42.4	46.8
France	9.9	11.7	10.8	9.6	13.7	15	15.7	18.2	23.6	29.8	33	38.4	39	45.5	47.6	46.5
Italy	8.1	14.8	19.4	18.5	25.7	25.1	24.7	24.6	18.3	11.9	7.3	6.1	7.9	8.7	9.1	10.3
UK	0.4	0.7	0.9	1.4	2.6	3.6	4.1	4.6	4.6	5.4	6.4	8.7	12.5	19	21.7	20.2
Sweden		6.7	6.2	5.4	4.1	2.2	1.3	1.3	0.9	0.6	0.6	0.9	0.9	3.1	3.2	2.8
Sales weighted average*	7.2	10.3	11	10	13.4	15.6	16.9	15.7	14.4	14	13.9	14.7	17.1	20	22.6	22.1

* All Europe for which data recorded (8 countries in 1980, 17 countries in 1995).
Source: ACEA/OICA, 1997.

Table D3. **Evolution of market segments in 7 countries European sales weighted average (%)**

	1980	1981	1982	1983	1984	1985	1986	1987	1988	1989	1990	1991	1992	1993	1994	1995
Superior and luxury range	18.1	17.1	16.2	16.2	15.7	15.7	15.4	14.9	15	15	14	14	13	14	14	15
Upper medium range	26	24	23	24	22	21	20	21	21	22	22	21	20	20	20	19
Lower medium range	28	30	31	30	30	29	29	29	29	29	28	29	31	32	30	30
Economical and smaller range	26.4	27.6	28	28.1	29.6	32.7	32	33	32	31	31	32	32	30	31	32
Other	2	2	2	2	2	2	3	3	3	3	5	4	4	4	4	4

Source: ACEA/OICA, 1997.

Table D4. **Sales weighted average power and engine displacement**

	1980	1981	1982	1983	1984	1985	1986	1987	1988	1989	1990	1991	1992	1993	1994	1995
Engine displacement (cc)	1 457	1 458	1 465	1 490	1 509	1 523	1 544	1 558	1 575	1 584	1 595	1 610	1 634	1 657	1 666	1 663
Power (kW)	51	51	52	54	54	55	55	57	58	59	61	61	62	63	64	65

Source: ACEA/OICA, 1997.

Table D5. **Sales weighted average power (kW)**

	1980	1981	1982	1983	1984	1985	1986	1987	1988	1989	1990	1991	1992	1993	1994	1995
Germany	59	58	59	61	61	61	60	62	64	67	68	66	68	69	69	70
Austria	56	56	57	58	57	57	57	55	59	61	62	65	65	64	64	64
Belgium	52	51	51	50	50	51	52	53	54	55	57	59	61	62	63	63
France	46	46	47	49	51	52	53	54	54	55	55	55	55	57	56	56
Italy	44	43	43	43	43	44	45	47	48	49	52	54	54	54	57	60
UK	54	54	54	56	56	58	59	60	61	63	64	65	65	66	66	68
Sweden		67	67	70	71	72	73	73	76	79	81	84	87	88	90	92

Source: ACEA/OICA, 1997.

Table D6. **Petrol and diesel pump prices - including excise taxes and VAT**

(1990 US $ per litre)

	LEADED PREMIUM PETROL								
	Austria	Belgium	France	Germany	Italy	Sweden	UK	OECD Europe	Ave. of Seven
1980	0.947	1.225	1.473	0.829	2.05	1.449	1.242	1.886	1.274
1981	0.872	1.068	1.155	0.764	1.66	1.291	1.141	1.524	1.099
1982	0.831	0.87	0.977	0.663	1.392	1.076	0.989	1.252	0.938
1983	0.748	0.762	0.828	0.602	1.231	0.856	0.883	1.041	0.822
1984	0.655	0.641	0.728	0.538	1.056	0.74	0.764	0.836	0.717
1985	0.64	0.631	0.724	0.523	0.912	0.733	0.738	0.755	0.686
1986	0.683	0.627	0.772	0.531	1.069	0.755	0.707	0.759	0.713
1987	0.792	0.718	0.887	0.623	1.192	0.825	0.766	0.821	0.804
1988	0.768	0.717	0.866	0.615	1.182	0.858	0.786	0.801	0.8
1989	0.754	0.72	0.84	0.673	1.068	0.823	0.724	0.769	0.781
1990	0.906	0.912	0.981	0.793	1.23	1.093	0.798	0.889	0.908
1991	0.815	0.89	0.918	0.837	1.161	1.025	0.809	0.877	0.901
1992	0.902	0.933	0.938	0.899	1.107	1.009	0.804	0.887	0.921
1993	n.a.	0.886	0.889	0.809	0.88	0.875	0.729	0.784	0.82
1994	n.a.	0.917	0.921	0.893	0.865	0.843	0.762	0.806	0.859
1995	n.a.	1.02	1.053	0.992	0.88	0.89	0.797	0.874	0.931
1996	n.a.	1.12	1.113	n.a.	0.906	n.a.	n.a.	0.951	1.018

Source: International Energy Agency database, Paris 1997.

Table D6. **Petrol and Diesel Pump Prices – including excise taxes and VAT** *(cont.)*

(1990 US $ per litre)

	AUTOMOTIVE DIESEL FUEL								
	Austria	Belgium	France	Germany	Italy	Sweden	UK	OECD Europe	Seven
1980	0.933	n.a.	1.206	0.81	0.913	0.72	1.251	1.012	1.016
1981	0.841	n.a.	0.995	0.689	0.734	0.661	1.131	0.878	0.873
1982	0.734	n.a.	0.729	0.624	0.681	0.636	0.967	0.747	0.742
1983	0.674	n.a.	0.632	0.557	0.633	0.522	0.842	0.655	0.656
1984	0.6	n.a.	0.557	0.502	0.54	0.508	0.722	0.576	0.577
1985	0.578	n.a.	0.554	0.485	0.5	0.483	0.719	0.558	0.563
1986	0.629	n.a.	0.55	0.49	0.514	0.424	0.675	0.552	0.554
1987	0.705	n.a.	0.607	0.551	0.603	0.506	0.7	0.609	0.61
1988	0.682	n.a.	0.575	0.527	0.624	0.509	0.714	0.598	0.602
1989	0.623	n.a.	0.546	0.516	0.624	0.554	0.649	0.569	0.576
1990	0.761	0.67	0.629	0.62	0.812	0.832	0.719	0.68	0.687
1991	0.707	0.677	0.604	0.621	0.853	0.783	0.729	0.681	0.69
1992	0.677	0.721	0.618	0.623	0.818	0.743	0.719	0.675	0.686
1993	0.623	0.668	0.6	0.57	0.67	0.579	0.663	0.613	0.619
1994	0.605	0.67	0.633	0.602	0.631	0.672	0.69	0.625	0.639
1995	0.728	0.736	0.691	0.657	0.645	0.716	0.723	0.671	0.684
1996	0.776	n.a.	0.769	n.a.	0.688	n.a.	n.a.	0.699	0.732

Source: International Energy Agency database, Paris 1997.

Table D7. **Average fuel consumption of new passenger cars weighted by registrations**
(litres per 100 km according to norm 80/1268/EEC)

	1993	1994	1995
GERMANY	7.60	7.48	7.42
AUSTRIA	7.15	7.03	6.98
BELGIUM	6.93	6.85	6.72
DENMARK	7.44	7.30	7.23
SPAIN	6.91	6.77	6.64
FRANCE	6.63	6.55	6.52
IRELAND	6.75	6.81	6.85
ITALY	6.94	6.95	6.97
LUXEMBOURG	7.39	7.39	7.38
NORWAY	7.45	7.56	7.56
NETHERLANDS	7.25	7.19	7.15
PORTUGAL	6.57	6.60	6.62
UNITED KINGDOM	7.26	7.24	7.26
SWEDEN	8.40	8.40	8.52
SWITZERLAND	8.43	8.36	8.30
AVERAGE	**7.19**	**7.12**	**7.10**

Source: CCFA/AAA, 1996.

◆ Figure D1. **Average fuel consumption of new passenger cars weighted by registrations**
(litres per 100 km according to norm 80/1268/EEC)

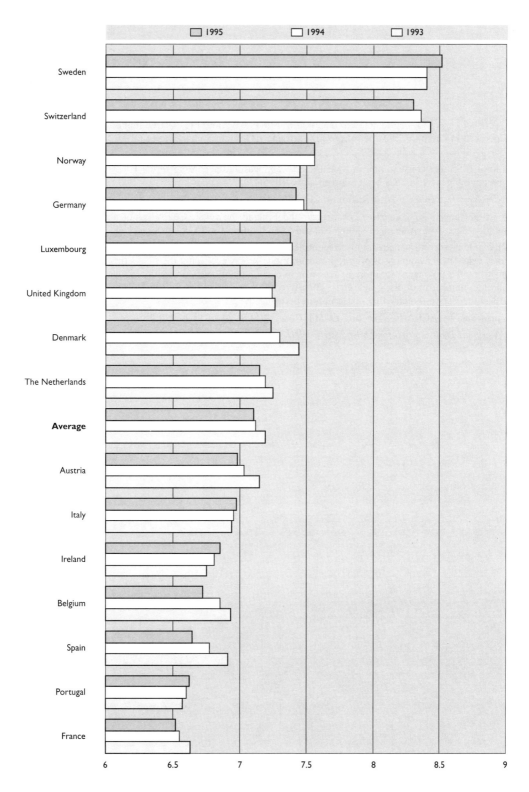

Source: CCFA/AAA, 1996.

Notes

1. Diesel produces less CO_2 per vehicle km than petrol if you take current respective fuel consumption averages of 6.5 litres diesel per 100 km and 8.0 litres petrol per 100 km and CO_2 emissions averages of 2.62 kg/l of diesel and 2.34 kg/l of petrol.
2. While until 1995 fuel consumption in Switzerland was measured according to the FTP 75 (Federal Test Procedure), the new Directive 93/116/EEC has been adopted since 1996.
3. *Hersteller-Nr / Typschlüssel-Nr. / Ausführungsschlüssel-Nr.*
4. CCFA in fact covers 17 countries. For Greece and Finland, however, CCFA is not able to link registration and fuel consumption data, so that average figures can only be produced for 15 countries.
5. The VIN code is a world-wide recognised identification number under UN conventions and can contain several characteristics of a vehicle (like country of origin, producer, model). However, only the first 3 (of the 17) positions in the code are binding while the producers are more or less free to encode additional information about the vehicle (characteristics like transmission, number of doors, gears, etc.). Therefore, there is a great heterogeneity in applying the code. In general, it is not possible to derive directly fuel consumption by using the VIN code only.
6. Table presented by ACEA in a meeting ECMT / ACEA / OICA on 12 June 1996.
7. VDA: *Verband Deutscher Automobilindustrie VDIK: Verband Deutscher Importeure von Kraftfahrzeugen.*

ANNEX

COUNCIL OF MINISTERS

DIALOGUE WITH VEHICLE MANUFACTURERS

At the ECMT Ministerial Session in Vienna on 8 June 1995 the Ministers of Transport held a dialogue with the vehicle manufacturing industry on the reduction of CO_2 emissions from passenger cars.

The meeting was chaired by Mr. Viktor Klima, Minister of Transport of Austria. Industry was represented by Mr. G. Garuzzo (President of ACEA and of Fiat Auto SpA), Mr. J.-Y. Helmer (Director of Automobile Division, PSA Peugeot-Citroen), Mr. H. Demel (President of the Board, Audi AG) and Mr. A. Diekmann (President of OICA).

Ministers and Industry agreed to work together to reduce substantially CO_2 emissions from new cars and to achieve tangible and steady reductions in in-use vehicle emissions. Government and Industry agreed on the following declaration at the conclusion of the Dialogue.

JOINT DECLARATION ON REDUCING CARBON DIOXIDE EMISSIONS FROM PASSENGER VEHICLES IN ECMT COUNTRIES

The Council of the ECMT and the Vehicle Manufacturing Industry (represented by OICA and ACEA), meeting in Vienna on 7-8 June, 1995, have agreed as follows:

1. Background

In the 1992 Framework Convention on Climate Change (FCCC), the Governments of industrialised countries agreed to work towards the stabilisation of greenhouse gas emissions at 1990 levels by the year 2000, and to reduce them thereafter. Although the Framework Convention does not specify these targets at the sectoral level, it is clear that the stabilisation of transport-based greenhouse gases in Europe will be required over the medium term. However, it is also recognised that the constraints of market demands and cost-effectiveness will affect the time frame over which such a goal can actually be realised.

Many different measures will be needed to reduce greenhouse gas emissions from the transport sector. With direct responsibility for this sector, Transport Ministers will have a key role to play in the design and implementation of these measures. Because automobiles are a major source of the transport sector's carbon dioxide emissions (the most important greenhouse gas), automobile manufacturers will also be expected to contribute significantly to the reduction of these emissions.

ECMT Ministers and the vehicle manufacturing industry therefore agree on the need for a joint approach to reducing CO_2 emissions from automobiles. A voluntary accord between Government and Industry is an important opportunity for each to express their fundamental interest in improving the CO_2 performance of automobile construction and use.

This Declaration is one step toward that long-term goal. In moving along this path, it is recognised that, in the early stages at least, the process of working together to achieve tangible progress may be more important than any quantified environmental target. This Declaration is intended to accelerate that co-operative process.

2. Objectives

The objectives of this Declaration are:
– to substantially and continuously reduce the fuel consumption of new cars sold in ECMT countries;
– to manage vehicle use so as to achieve tangible and steady reductions in their total CO_2 emissions.

A number of governments have already introduced, or are considering introducing, CO_2 targets for the transport sector. Some are negotiating with Industry. This Declaration is not intended to limit the scope for such initiatives.

It is also recognised that fuel economy is becoming a competitive issue within industry, though the starting points differ from country to country. Though this will influence fuel economy further action will be required on the part of both Government and Industry if these joint objectives are to be achieved. In some instances, the primary initiative should be taken by Industry, with support from Government. In others, the reverse will be true. Although the degree of responsibility will vary according to the measure involved, each of these groups will have some role to play in the success of all measures.

3. Measures

3.1 Government measures

Policy framework

Governments will set the broad policy framework for the transport system. This policy should be economically-efficient and take full account of all environmental impacts.

Government will continue to use economic instruments, environmental regulations, information and other measures to influence the market for, and to encourage the use of, fuel-efficient vehicles in a safe, fuel-conserving, manner. Government will also strive to ensure that policy measures taken in related vehicle design areas (e.g. safety and noise) are consistent with the need to reduce greenhouse gas emissions.

In implementing the above commitments, Government will apply the following general principles:

a) Measures taken will be implemented in as cost-effective a manner as practical.
b) Government accepts that the demand for more fuel-efficient cars should come essentially from the consumer, operating in free markets. Government policies will therefore strive to encourage consumers to choose fuel-efficient vehicles, and to operate them in a fuel-efficient way.
c) Government will seek to avoid major disruptions in policy, aiming instead for gradual, steady, and consistent implementation, so as to decrease uncertainty in the marketplace.
d) Irrespective of the type of measure being employed, international co-ordination will be pursued, to help avoid discrimination among individual countries or firms, and to provide a coherent message to Industry about future policy directions in Europe as a whole.
e) Government will actively consult with Industry on all significant policy initiatives taken in the pursuit of the goals contained in this Declaration.

New technologies and road traffic informatics

Government will encourage the creation and introduction of new information technologies, where they can provide a cost-effective means to reduce congestion and related losses in fuel consumption. The potential of integrated traffic management systems will be given special attention (e.g. increased use of public transport, combined with controlled access to city centres; road information/guidance systems; appropriate infrastructural measures).

Fleet maintenance/replacement

Government undertakes to develop and introduce harmonized systems of regular vehicle inspection and maintenance, in order to make the existing automobile fleet as clean and fuel-efficient as possible. Government also undertakes to investigate cost-effective ways of encouraging the disposal/scrappage of the oldest, dirtiest and most fuel-inefficient vehicles, provided that this would improve total global emissions, calculated on a full-fuel-cycle basis.

3.2 Industry measures

Fuel-efficient new cars

Industry agrees to give a strong emphasis to developing, manufacturing and marketing vehicles with improved fuel efficiencies. As the owners and developers of car manufacturing technologies, Industry is well-placed to promote the incorporation of new, fuel-efficient, techniques into vehicle designs, so as to continuously and significantly improve the fuel consumption profile of the fleet.

Marketing

Industry undertakes to promote energy efficiency as a sales argument. Conversely, the concepts of power, acceleration, and maximum speed will not be used as major sales arguments.

Industry recognises that it occupies a special place in the transport marketplace, and therefore has a special duty to demonstrate to consumers how its vehicles can be used in an environmentally-responsible manner. With regard to fuel efficiency, Industry will explicitly examine the idea of an advertising "code of practice".

3.3 Joint Government and Industry measures

Marketing

Based on existing EC criteria (Directive 93/116/EC) for measuring CO_2 emissions/fuel consumption, Industry and Government undertake to examine the possibilities of, and if appropriate, to define practical arrangements for, introducing a standardised labelling system for new cars.

Developing new technologies

Information technologies (telematics) often require new kinds of equipment for vehicles. Manufacturers and governments will co-operate closely to define the criteria such equipment should meet, as well as to introduce them in practice. For example, ERTICO, where some Governments and Industry are already both represented, provides one valuable mechanism for ensuring that this technology can be applied efficiently.

Research and development

Government and Industry agree that more emphasis needs to be placed on improving R&D programmes related to CO_2 emissions from cars. Both therefore undertake to work toward better co-ordination of existing R&D efforts, especially at the European level.

Because technology development is so crucial to future fuel efficiencies, all reasonable opportunities to encourage joint R&D programmes between Industry and Government should be fully explored. Existing R&D programmes of the European Union or of Industry (e.g. EUCAR), as well as the International Energy Agency's Implementing Agreements related to research and development should all be exploited in this context. Both basic research and its uptake in the marketplace will be emphasized in these activities.

Information/education

Specialised information should be developed for vehicle users, vehicle dealers and importers, and driving instructors, in order to promote fuel efficiency with regard to car purchase, use and driver behaviour. Government and Industry agree to develop specialised education/information campaigns aimed at these individual publics.

Other initiatives

Government and Industry agree to study the environmental value and economic feasibility of further consumer-oriented initiatives that would help to improved driving style and fuel consumption as well as traffic management, including, for example, econometers or on-board computers to indicate fuel consumption, the relationship between power, speed capability and fuel economy, and fuel-conserving traffic management measures.

4. Monitoring

Governments and Industry agree to establish an appropriate system to monitor progress toward the goals contained in this Declaration. This monitoring system should:

    a) Analyse trends in the projected specific fuel consumption and/or CO_2 emissions of new cars sold in ECMT Member countries.
        Beginning in 1996, data will be provided annually for each ECMT Member country, and in a standardized form, on:
            i) the number of new car registrations, making appropriate distinctions among vehicle characteristics; and

ii) specific fuel consumption (in litres/100 kilometres) and/or CO_2 emissions (in grammes per kilometre) of these cars.

b) Periodically assess the effectiveness and efficiency of measures taken by both Government and Industry towards achieving the objectives of this Declaration.

Beginning in 1997, and continuing biannually thereafter, Government and Industry (or the Industry Associations) will report on all measures taken in support of this Declaration, including a qualitative evaluation of the effectiveness of these measures.

c) Periodically review the objectives of this Declaration, in the light of future developments in the international debate concerning climate change.

MAIN SALES OUTLETS OF OECD PUBLICATIONS
PRINCIPAUX POINTS DE VENTE DES PUBLICATIONS DE L'OCDE

AUSTRALIA – AUSTRALIE
D.A. Information Services
648 Whitehorse Road, P.O.B 163
Mitcham, Victoria 3132　　Tel. (03) 9210.7777
　　　　　　　　　　　　　Fax: (03) 9210.7788

AUSTRIA – AUTRICHE
Gerold & Co.
Graben 31
Wien I　　　　　　　　　Tel. (0222) 533.50.14
　　　　　　　　　　　　Fax: (0222) 512.47.31.29

BELGIUM – BELGIQUE
Jean De Lannoy
Avenue du Roi, Koningslaan 202
B-1060 Bruxelles　　Tel. (02) 538.51.69/538.08.41
　　　　　　　　　　　　Fax: (02) 538.08.41

CANADA
Renouf Publishing Company Ltd.
1294 Algoma Road
Ottawa, ON K1B 3W8　　　Tel. (613) 741.4333
　　　　　　　　　　　　Fax: (613) 741.5439
Stores:
61 Sparks Street
Ottawa, ON K1P 5R1　　　Tel. (613) 238.8985

12 Adelaide Street West
Toronto, ON M5H 1L6　　Tel. (416) 363.3171
　　　　　　　　　　　　Fax: (416)363.59.63

Les Éditions La Liberté Inc.
3020 Chemin Sainte-Foy
Sainte-Foy, PQ G1X 3V6　　Tel. (418) 658.3763
　　　　　　　　　　　　Fax: (418) 658.3763

Federal Publications Inc.
165 University Avenue, Suite 701
Toronto, ON M5H 3B8　　Tel. (416) 860.1611
　　　　　　　　　　　　Fax: (416) 860.1608

Les Publications Fédérales
1185 Université
Montréal, QC H3B 3A7　　Tel. (514) 954.1633
　　　　　　　　　　　　Fax: (514) 954.1635

CHINA – CHINE
China National Publications Import
Export Corporation (CNPIEC)
16 Gongti E. Road, Chaoyang District
P.O. Box 88 or 50
Beijing 100704 PR　　　　Tel. (01) 506.6688
　　　　　　　　　　　　Fax: (01) 506.3101

CHINESE TAIPEI – TAIPEI CHINOIS
Good Faith Worldwide Int'l. Co. Ltd.
9th Floor, No. 118, Sec. 2
Chung Hsiao E. Road
Taipei　　　　　　Tel. (02) 391.7396/391.7397
　　　　　　　　　　　　Fax: (02) 394.9176

CZECH REPUBLIC – RÉPUBLIQUE TCHÈQUE
National Information Centre
NIS – prodejna
Konviktská 5
Praha 1 – 113 57　　　　Tel. (02) 24.23.09.07
　　　　　　　　　　　　Fax: (02) 24.22.94.33
(*Contact* Ms Jana Pospisilova, nkposp@dec.niz.cz)

DENMARK – DANEMARK
Munksgaard Book and Subscription Service
35, Nørre Søgade, P.O. Box 2148
DK-1016 København K　　Tel. (33) 12.85.70
　　　　　　　　　　　　Fax: (33) 12.93.87

J. H. Schultz Information A/S,
Herstedvang 12,
DK – 2620 Albertslung　　Tel. 43 63 23 00
　　　　　　　　　　　　Fax: 43 63 19 69
Internet: s-info@inet.uni-c.dk

EGYPT – ÉGYPTE
The Middle East Observer
41 Sherif Street
Cairo　　　　　　　　　Tel. 392.6919
　　　　　　　　　　　　Fax: 360-6804

FINLAND – FINLANDE
Akateeminen Kirjakauppa
Keskuskatu 1, P.O. Box 128
00100 Helsinki

Subscription Services/Agence d'abonnements :
P.O. Box 23
00371 Helsinki　　　　　Tel. (358 0) 121 4416
　　　　　　　　　　　　Fax: (358 0) 121.4450

FRANCE
OECD/OCDE
Mail Orders/Commandes par correspondance :
2, rue André-Pascal
75775 Paris Cedex 16　　Tel. (33-1) 45.24.82.00
　　　　　　　　　　　Fax: (33-1) 49.10.42.76
　　　　　　　　　　　Telex: 640048 OCDE
Internet: Compte.PUBSINQ@oecd.org

Orders via Minitel, France only/
Commandes par Minitel, France exclusivement :
36 15 OCDE

OECD Bookshop/Librairie de l'OCDE :
33, rue Octave-Feuillet
75016 Paris　　　　　　Tél. (33-1) 45.24.81.81
　　　　　　　　　　　　(33-1) 45.24.81.67

Dawson
B.P. 40
91121 Palaiseau Cedex　　Tel. 69.10.47.00
　　　　　　　　　　　　Fax: 64.54.83.26

Documentation Française
29, quai Voltaire
75007 Paris　　　　　　Tel. 40.15.70.00

Economica
49, rue Héricart
75015 Paris　　　　　　Tel. 45.75.05.67
　　　　　　　　　　　　Fax: 40.58.15.70

Gibert Jeune (Droit-Économie)
6, place Saint-Michel
75006 Paris　　　　　　Tel. 43.25.91.19

Librairie du Commerce International
10, avenue d'Iéna
75016 Paris　　　　　　Tel. 40.73.34.60

Librairie Dunod
Université Paris-Dauphine
Place du Maréchal-de-Lattre-de-Tassigny
75016 Paris　　　　　　Tel. 44.05.40.13

Librairie Lavoisier
11, rue Lavoisier
75008 Paris　　　　　　Tel. 42.65.39.95

Librairie des Sciences Politiques
30, rue Saint-Guillaume
75007 Paris　　　　　　Tel. 45.48.36.02

P.U.F.
49, boulevard Saint-Michel
75005 Paris　　　　　　Tel. 43.25.83.40

Librairie de l'Université
12a, rue Nazareth
13100 Aix-en-Provence　　Tel. (16) 42.26.18.08

Documentation Française
165, rue Garibaldi
69003 Lyon　　　　　　Tel. (16) 78.63.32.23

Librairie Decitre
29, place Bellecour
69002 Lyon　　　　　　Tel. (16) 72.40.54.54

Librairie Sauramps
Le Triangle
34967 Montpellier Cedex 2　Tel. (16) 67.58.85.15
　　　　　　　　　　　　Fax: (16) 67.58.27.36

A la Sorbonne Actual
23, rue de l'Hôtel-des-Postes
06000 Nice　　　　　　Tel. (16) 93.13.77.75
　　　　　　　　　　　　Fax: (16) 93.80.75.69

GERMANY – ALLEMAGNE
OECD Bonn Centre
August-Bebel-Allee 6
D-53175 Bonn　　　　　Tel. (0228) 959.120
　　　　　　　　　　　Fax: (0228) 959.12.17

GREECE – GRÈCE
Librairie Kauffmann
Stadiou 28
10564 Athens　　　　　Tel. (01) 32.55.321
　　　　　　　　　　　Fax: (01) 32.30.320

HONG-KONG
Swindon Book Co. Ltd.
Astoria Bldg. 3F
34 Ashley Road, Tsimshatsui
Kowloon, Hong Kong　　Tel. 2376.2062
　　　　　　　　　　　Fax: 2376.0685

HUNGARY – HONGRIE
Euro Info Service
Margitsziget, Európa Ház
1138 Budapest　　　　Tel. (1) 111.62.16
　　　　　　　　　　　Fax: (1) 111.60.61

ICELAND – ISLANDE
Mál Mog Menning
Laugavegi 18, Pósthólf 392
121 Reykjavik　　　　Tel. (1) 552.4240
　　　　　　　　　　　Fax: (1) 562.3523

INDIA – INDE
Oxford Book and Stationery Co.
Scindia House
New Delhi 110001　　Tel. (11) 331.5896/5308
　　　　　　　　　　　Fax: (11) 371.8275

17 Park Street
Calcutta 700016　　　　Tel. 240832

INDONESIA – INDONÉSIE
Pdii-Lipi
P.O. Box 4298
Jakarta 12042　　　　Tel. (21) 573.34.67
　　　　　　　　　　　Fax: (21) 573.34.67

IRELAND – IRLANDE
Government Supplies Agency
Publications Section
4/5 Harcourt Road
Dublin 2　　　　　　　Tel. 661.31.11
　　　　　　　　　　　Fax: 475.27.60

ISRAEL – ISRAËL
Praedicta
5 Shatner Street
P.O. Box 34030
Jerusalem 91430　　　Tel. (2) 52.84.90/1/2
　　　　　　　　　　　Fax: (2) 52.84.93

R.O.Y. International
P.O. Box 13056
Tel Aviv 61130　　　　Tel. (3) 546 1423
　　　　　　　　　　　Fax: (3) 546 1442

Palestinian Authority/Middle East:
INDEX Information Services
P.O.B. 19502
Jerusalem　　　　　　Tel. (2) 27.12.19
　　　　　　　　　　　Fax: (2) 27.16.34

ITALY – ITALIE
Libreria Commissionaria Sansoni
Via Duca di Calabria 1/1
50125 Firenze　　　　Tel. (055) 64.54.15
　　　　　　　　　　　Fax: (055) 64.12.57

Via Bartolini 29
20155 Milano　　　　Tel. (02) 36.50.83

Editrice e Libreria Herder
Piazza Montecitorio 120
00186 Roma　　　　　Tel. 679.46.28
　　　　　　　　　　　Fax: 678.47.51

Libreria Hoepli
Via Hoepli 5
20121 Milano　　　　Tel. (02) 86.54.46
　　　　　　　　　　　Fax: (02) 805.28.86

Libreria Scientifica
Dott. Lucio de Biasio 'Aeiou'
Via Coronelli, 6
20146 Milano　　　　Tel. (02) 48.95.45.52
　　　　　　　　　　　Fax: (02) 48.95.45.48

JAPAN – JAPON
OECD Tokyo Centre
Landic Akasaka Building
2-3-4 Akasaka, Minato-ku
Tokyo 107 Tel. (81.3) 3586.2016
 Fax: (81.3) 3584.7929

KOREA – CORÉE
Kyobo Book Centre Co. Ltd.
P.O. Box 1658, Kwang Hwa Moon
Seoul Tel. 730.78.91
 Fax: 735.00.30

MALAYSIA – MALAISIE
University of Malaya Bookshop
University of Malaya
P.O. Box 1127, Jalan Pantai Baru
59700 Kuala Lumpur
Malaysia Tel. 756.5000/756.5425
 Fax: 756.3246

MEXICO – MEXIQUE
OECD Mexico Centre
Edificio INFOTEC
Av. San Fernando no. 37
Col. Toriello Guerra
Tlalpan C.P. 14050
Mexico D.F. Tel. (525) 665 47 99
 Fax: (525) 606 13 07

NETHERLANDS – PAYS-BAS
SDU Uitgeverij Plantijnstraat
Externe Fondsen
Postbus 20014
2500 EA's-Gravenhage Tel. (070) 37.89.880
Voor bestellingen: Fax: (070) 34.75.778

Subscription Agency/
Agence d'abonnements :
SWETS & ZEITLINGER BV
Heereweg 347B
P.O. Box 830
2160 SZ Lisse Tel. 252.435.111
 Fax: 252.415.888

NEW ZEALAND – NOUVELLE-ZÉLANDE
GPLegislation Services
P.O. Box 12418
Thorndon, Wellington Tel. (04) 496.5655
 Fax: (04) 496.5698

NORWAY – NORVÈGE
NIC INFO A/S
Ostensjoveien 18
P.O. Box 6512 Etterstad
0606 Oslo Tel. (22) 97.45.00
 Fax: (22) 97.45.45

PAKISTAN
Mirza Book Agency
65 Shahrah Quaid-E-Azam
Lahore 54000 Tel. (42) 735.36.01
 Fax: (42) 576.37.14

PHILIPPINE – PHILIPPINES
International Booksource Center Inc.
Rm 179/920 Cityland 10 Condo Tower 2
HV dela Costa Ext cor Valero St.
Makati Metro Manila Tel. (632) 817 9676
 Fax: (632) 817 1741

POLAND – POLOGNE
Ars Polona
00-950 Warszawa
Krakowskie Prezdmiescie 7 Tel. (22) 264760
 Fax: (22) 265334

PORTUGAL
Livraria Portugal
Rua do Carmo 70-74
Apart. 2681
1200 Lisboa Tel. (01) 347.49.82/5
 Fax: (01) 347.02.64

SINGAPORE – SINGAPOUR
Ashgate Publishing
Asia Pacific Pte. Ltd
Golden Wheel Building, 04-03
41, Kallang Pudding Road
Singapore 349316 Tel. 741.5166
 Fax: 742.9356

SPAIN – ESPAGNE
Mundi-Prensa Libros S.A.
Castelló 37, Apartado 1223
Madrid 28001 Tel. (91) 431.33.99
 Fax: (91) 575.39.98

Mundi-Prensa Barcelona
Consell de Cent No. 391
08009 – Barcelona Tel. (93) 488.34.92
 Fax: (93) 487.76.59

Llibreria de la Generalitat
Palau Moja
Rambla dels Estudis, 118
08002 – Barcelona
 (Subscripcions) Tel. (93) 318.80.12
 (Publicacions) Tel. (93) 302.67.23
 Fax: (93) 412.18.54

SRI LANKA
Centre for Policy Research
c/o Colombo Agencies Ltd.
No. 300-304, Galle Road
Colombo 3 Tel. (1) 574240, 573551-2
 Fax: (1) 575394, 510711

SWEDEN – SUÈDE
CE Fritzes AB
S–106 47 Stockholm Tel. (08) 690.90.90
 Fax: (08) 20.50.21

For electronic publications only/
Publications électroniques seulement
STATISTICS SWEDEN
Informationsservice
S-115 81 Stockholm Tel. 8 783 5066
 Fax: 8 783 4045

Subscription Agency/Agence d'abonnements :
Wennergren-Williams Info AB
P.O. Box 1305
171 25 Solna Tel. (08) 705.97.50
 Fax: (08) 27.00.71

SWITZERLAND – SUISSE
Maditec S.A. (Books and Periodicals/Livres et périodiques)
Chemin des Palettes 4
Case postale 266
1020 Renens VD 1 Tel. (021) 635.08.65
 Fax: (021) 635.07.80

Librairie Payot S.A.
4, place Pépinet
CP 3212
1002 Lausanne Tel. (021) 320.25.11
 Fax: (021) 320.25.14

Librairie Unilivres
6, rue de Candolle
1205 Genève Tel. (022) 320.26.23
 Fax: (022) 329.73.18

Subscription Agency/Agence d'abonnements :
Dynapresse Marketing S.A.
38, avenue Vibert
1227 Carouge Tel. (022) 308.08.70
 Fax: (022) 308.07.99

See also – Voir aussi :
OECD Bonn Centre
August-Bebel-Allee 6
D-53175 Bonn (Germany) Tel. (0228) 959.120
 Fax: (0228) 959.12.17

THAILAND – THAÏLANDE
Suksit Siam Co. Ltd.
113, 115 Fuang Nakhon Rd.
Opp. Wat Rajbopith
Bangkok 10200 Tel. (662) 225.9531/2
 Fax: (662) 222.5188

TRINIDAD & TOBAGO, CARIBBEAN
TRINITÉ-ET-TOBAGO, CARAÏBES
SSL Systematics Studies Limited
9 Watts Street
Curepe
Trinadad & Tobago, W.I. Tel. (1809) 645.3475
 Fax: (1809) 662.5654

TUNISIA – TUNISIE
Grande Librairie Spécialisée
Fendri Ali
Avenue Haffouz Imm El-Intilaka
Bloc B 1 Sfax 3000 Tel. (216-4) 296 855
 Fax: (216-4) 298.270

TURKEY – TURQUIE
Kültür Yayinlari Is-Türk Ltd. Sti.
Atatürk Bulvari No. 191/Kat 13
06684 Kavaklidere/Ankara
 Tél. (312) 428.11.40 Ext. 2458
 Fax : (312) 417.24.90
 et 425.07.50-51-52-53

Dolmabahce Cad. No. 29
Besiktas/Istanbul Tel. (212) 260 7188

UNITED KINGDOM – ROYAUME-UNI
HMSO
Gen. enquiries Tel. (0171) 873 0011
Postal orders only:
P.O. Box 276, London SW8 5DT
Personal Callers HMSO Bookshop
49 High Holborn, London WC1V 6HB
 Fax: (0171) 873 8463

Branches at: Belfast, Birmingham, Bristol, Edinburgh, Manchester

UNITED STATES – ÉTATS-UNIS
OECD Washington Center
2001 L Street N.W., Suite 650
Washington, D.C. 20036-4922 Tel. (202) 785.6323
 Fax: (202) 785.0350
Internet: washcont@oecd.org

Subscriptions to OECD periodicals may also be placed through main subscription agencies.

Les abonnements aux publications périodiques de l'OCDE peuvent être souscrits auprès des principales agences d'abonnement.

Orders and inquiries from countries where Distributors have not yet been appointed should be sent to: OECD Publications, 2, rue André-Pascal, 75775 Paris Cedex 16, France.

Les commandes provenant de pays où l'OCDE n'a pas encore désigné de distributeur peuvent être adressées aux Éditions de l'OCDE, 2, rue André-Pascal, 75775 Paris Cedex 16, France.

8-1996